For my old buddy and
severest critic! — Darrell Dabney

— Dale Riepe

OBJECTIVITY AND SUBJECTIVISM IN THE PHILOSOPHY OF SCIENCE WITH SPECIAL REFERENCE TO INDIA

OBJECTIVITY AND SUBJECTIVISM IN THE PHILOSOPHY OF SCIENCE WITH SPECIAL REFERENCE TO INDIA

Dale Riepe
Professor of Philosophy
State University of New York at Buffalo

K P BAGCHI & COMPANY
CALCUTTA • NEW DELHI

First Published in January 1986

K P BAGCHI & COMPANY
286 BB Ganguli Street, Calcutta 700012
J-1698 Chittaranjan Park, New Delhi 110019

Published by

K K Bagchi
K P Bagchi & Co
286 BB Ganguli Street,
Calcutta 700012

Printed at Cambridge Printing Works, B-85 Naraina Industrial Area
Phase II, New Delhi □ Phone 587039

DEBIPRASAD CHATTOPADHYAYA, D. LITT., SC.D,
FRIEND OF TWENTY-FIVE YEARS
HISTORIAN OF INDIAN PHILOSOPHY

ACKNOWLEDGEMENTS

Sincere thanks go to Roland Grüner of Grüner Publishing Company in Amsterdam and to David De Grood and Ralph M. Faris for allowing the author to reprint, with revisions, "Contemporary Western Subjectivism in the Philosophy of Science," which first appeared in *Crisis and Consciousness* in 1977. To Debiprasad Chattopadhyaya goes the author's multi-faceted indebtedness for allowing the author to reprint with considerable revision and additions "Science and Contemporary Philosophy of Science in India, "*History and Society,* Essays in Honour of Professor Niharranjan Ray published by P.C. Bagchi in Calcutta. Debiprasad and De Grood together have been invaluable and constant critics and editors for more than twenty years.

Jerry Drost, research and reference librarian of the Lockwood Memorial Library, deserves special recognition for his help and Christoper Pines, doctoral candidate in philosophy at Buffalo, deserves thanks for his suggestions and assistance. Marie Fleischauer and Judy Wagner of my department were most helpful in preparing the manuscript.

PREFACE

When I first began doing research for this volume I got much assistance from Satyavan P. Kanal, Professor of Philosophy at the University of Delhi and from his colleague Professor R.S. Gill who introduced the topic of Philosophy of Science into the Delhi University curriculum and at other institutions in the Panjab. He brought to the study of science not only scientific theory but the practice of a railroad engineer (not an engine driver) in Burma many years ago. Kanal has been of inestimable help to me over the past fifteen years in keeping me informed of recent developments in Indian thought and in humanitarian progress since we first took tea together. The great benefactor of the Indian working class in Madras for many years, Dr. S. Devasagayan was not only my personal physician in South India, but helped me see the necessary relationship between scientific progress and improvements in civilization's upward path during our Saturday night walks through the chawls of a great city. Professor Abdur Rahman, since we first met at a scientific conference in New York in 1966, has kept me informed about the social implications of science which he has called to my attention with unfailing eloquence. In this respect, he has been doing for science what Debiprasad Chattopadhyaya has been doing for philosophy. Each of these great-souled Indians has dedicated his life to social progress in a constructive yet considerably different way.

Calcutta 25 March, 1984 — Dale Riepe

FOREWORD

India is faced with the twin problems of incorporating values of science and a scientific outlook as a part of her culture, including those elements which are necessitated by increasing technological complexion of society. Secondly, she has also to see that in doing so she is able to distinguish between the grain and the chaff, *i.e.* able to discard the socio-economic and cultural frills incorporated in the scientific and technological tradition of Europe. Even though some of the latter have come to be regarded as an essential part of science and technology of European culture area and the prevalent economic system, since the Industrial Revolution.

In order to incorporate science and technology in contemporary Indian culture, it would be necessary to fully understand the historical and social context of science as well as the inherited culture. Further, the cultural traditions, including philosophies have to be critically analysed to find out their relevance or otherwise, to contemporary problems. This is particularly difficult in view of the:

(*i*) tendency to uncritically accept medieval ideas in order to distinguish Indian thought and culture from Europe and assert national features;

(*ii*) lack of appreciation of the context of medieval concepts and philosophies and their inapplicability to contemporary problems, which are a product of different socio-economic structure and political framework;
Further, also due to lack of comprehension of science as a powerful intellectual tool in analysing contemporary problems including those of society and its culture and providing alternatives which could meet a desired set of social and political goals;

(*iii*) high pressure salesmanship of medieval philosophies, cultural dimensions, ideas and attitudes, for political purposes, including from overseas.

Science and technology, in their present form, were implanted in India by the British to intensify their colonial gains. In order to camouflage their economic and political objectives, they emphasized only the technical dimensions of science and technology and built up a number of myths about science and technology and scientists to suit their purposes. In this lore, the universalism of science, all round benefits to culture and civilization were over-emphasized. However, while the facts of science are valid everywhere, the constructs and the purposes for which it is used are part of the culture and philosophy of society and its economic and political structure. It was the uncritical acceptance of some of these myths and ignoring of the socio-economic features which has led to the unmitigated problems of pollution, degradation of the environment, creation of aggressive, self-centred individualism, (under the guise of autonomy of the individual). These have led to deprivation and alienation of a large number, in fact majority of individuals. Also this led to resources being diverted to search for the solutions aimed at meeting individual's need, depending upon those who can pay rather than finding social solutions. As a result, Indian society faced a series of crises. However, the full nature and character of the crises could not be comprehended, since each dimension was compartmentalized. This crisis is being characterized also as a crisis of science, technology or materialism depending upon the context of the discussion.

Indian scientists, philosophers, instead of looking at the problems at some depth and analysing the situation in its multi-faceted dimensions, find it easier, when it is suggested to them, to accept such formulations as: *science requires to be spiritualized* or old wisdom of India can help to find a solution to the present crisis of modern society. Such suggestions from advanced Europe, and even more advanced United States, feed the national vanity, while detracting from a critical understanding of socio-political and economic dimensions of science and technology on the one hand and social evolution on the other. They also promote derivative thinking.

What is required, however, is the use of science to analyse the development of science and technology in Europe and America, understand its linkages with other sub-systems, political, economic, social etc. and different strata of society and apply the knowledge in understanding the nature of the contemporary crisis and to avoid the pitfalls for the future. For this, one needs a philosophy, which is in harmony with contemporary knowledge and can be applied to the realities of the situation.

Such a philosophical outlook has not been forthcoming. As a result one notices that problems and questions posed in the context of the situation in Europe and America are picked up as they come up and are discussed and attempts are made to apply them outside of historical and social context in India. Instead using the philosophical thought or conceptual framework of Europe, the medieval India philosophical framework is used (to give a semblance of distinctiveness and originality). As a result we get involved in non-problems with wrong methodologies.

To give an example, Indian Government established a National Commission for History of Science over thirty years ago, yet it has not made an impact. The members of the Commission (scientists of eminence) instead of using scientific methodology and techniques, to understand the nature and character of science in India in different periods of Indian history, in the context of its economic and political structure, goals, social organization and culture, tried to look for ideas, methods and techniques, themes etc. which could match the achievements subject/ideawise of Europe in the corresponding periods. As a result, a few books which were chronological records of innovations, new ideas etc. were published. These hardly provided any insights into the problems of scientific and technological developments and their special features in India. Further, in the absence of latter, history of science could not be developed as a discipline. What was taught was a repetition of only what was given in European or American books, making science and technology as a European phenomenon.

Professor Dale Riepe has filled a long felt gap, and done a valuable service in bringing to the surface the derivative character of recent philosophical thought in India and indicating its linkages with European-American philosophies, and hinting at possible reasons for such a linkage. His present publication, like his four earlier publications on India is a rich contribution and it is hoped that it would help people to have a second look at their philosophies.

March 1984

A. Rahman,
Director, Institute of Science,
Technology and Development
Council for Scientific and Industrial Research,
Government of India New Delhi

CONTENTS

I

SCIENCE IN INDIA

HISTORICAL INTRODUCTION

Much important scientific activity took place in India before the British Raj. There are accounts of this earlier development in a number of sources demonstrating activity from vedic times. The oldest text that I possess is *The Āryabhaṭīya* attributed to the fifth century. It contains a mathematical section, reflections on the elements of astronomical chronology, and comments on celestial spheres. Included is a statement of the realization that the daily rotation of the heavens is only apparent because it is due to the rotation of the earth around its axis.[1] Aristarchus (fl. ca. 281 B.C.), who was attached to the Alexandrian Museum, believed in the rotation of the earth on its axis and revolution around the sun as a fixed center. Joseph Needham in his account of Chinese astronomy in ancient times does not record an early view of rotation, although the Chinese were advanced in instrumentation and model construction[2].

There are a number of treatises on vedic mathematics, but it is difficult to know when they were composed. Presumably, these could be older than *The Āryabhaṭīya.*[3] George Sarton has given due attention to the mathematical and astronomical discoveries of the Indian in his encyclopedic *Introduction to the History of Science* in five volumes. Since publication of his work, considerable has been published, including *Studies in the History of Science in India* in two volumes (New Delhi, 1982) by Debiprasad Chattopadhyaya. Of his work we shall learn more later. In this work which is a compendium of the history of Indian science, we find the work of J. Filliozat on Indian astronomy (1955), Chin Keh-mu on the scientific exchange between India and China (1958), M. Asimov's work on India and Central Asia (1968), Abdur Rahman's research on science and technology in medieval India (1982), Irfan Habib's article on technological change and society in the thirteenth and fourteenth century (1969), Debiprasad Chattopadhyaya on the relation of philosophy, religion and medicine (1979) as well as I-Tsing's observations on medicine in India and China (2 d.c.).

Also famous for its antiquity is the Caraka-Saṃhitā, the composition of which could have been from the sixth century B.C. to the first century A.D. or even later. It is a treatise on medicine and became the source of much later treatment of Indian medicine[4].

If we shift our attention from what is sometimes seen as generally theoretical and move into technology, we find that considerable effort has been expended to reconstruct also the state of technology and mechanical arts in India since the early peasant communities of Western India. Although nothing comparable to the Oxford *History of Technology* (1954-58) has yet appeared on Indian technology, the material for such a treatise is gradually and painstakingly being assembled. With respect to warfare, for example, it is believed that the early warriors of the Indus Valley (from before 2500 B.C.) fought on foot with dirks, sword, knives, daggers, axes, spear, lance, mace and bow and arrows (the latter tipped with copper heads).[5] We learn from the Ṛgveda (1600-1000 B.C.) that chariots were employed in open battle fields. A chariot, an elephant, three horses, and five foot-soldiers formed a *patti*, the ideal atomic unit of the fighting force in epic times. Forts were constructed as early as 2500 B.C. with brock lookout towers appearing shortly thereafter.

Ancient shipbuilding seems to be more difficult to ascertain, but there are treatises in Sanskrit and Pāli referring to it while in the *Vṛiksh-Āyurveda* (1000-500 B.C.) refers to the kinds of wood to be effectively used in the construction of ships. An early authority on shipbuilding warned against joining together the planks of bottoms intended to be sea-going vessels because the iron could expose them to the influence of magnetic rocks bringing them within a magnetic field.[6] One almost hopes that they had premonitions of degaussing. Evidently at about the time of the writing of the *Rāmāyana* (a love epic about Rāma and Sita) there were ships that could hold from 500to 800 people.[7] The first representation of a ship on an Indian coin is dated at 3000 B.C. and found at Mohenjo Daro (about an hour's flight into the desert from Karachi). It is comparable in India to Babylon.

The need for constant reassessment will presumably never cease, but what now is required for more understanding of the history of Indian science is a broader account of its interconnections with social science and the movement of society as a whole. Besides an account of theoretical progress, we need more work on its interlacing with technology, mechanical devices, alchemy-chemistry, botany and geography. Some of this could be based on P. Ray's *History of Chemistry in Ancient and Medieval India* (1956).[8] Debiprasad's *Science and Society in Ancient India* (1976) emphasizes the relationship between medicine and social institutions in connection with the various philosophical viewpoints represented during that period. Suggestions as to how to proceed in the light of the science of society is indicated by the work of A. Rahman since the 1960's and by his important Bibliography of *Source Material of History of Science in Medieval India* (New Delhi, 1975) and his encyclopedic *Science and Technology in Medieval India* (Sanskrit, Arabic, Persian) in (New Delhi, 1982). Debiprasad has also done great service in publishing *Studies in the History of Science in India* in two volumes (New Delhi, 1982). Here in two volumes are thirty-eight crucial and seminal papers on Indian science. They range chronologically from Al-Bīrūnī (c. 1060) to Rahman (1982) or over 900 years of important texts for understanding the science of India. Of great interest is Debiprasad's "Introduction" in the first volume which raises important questions such as the reason for Greco-Roman "creeping paralysis" which indicated the impossibility of the classical Western world to move ahead scientifically from around the second century B.C. Farrington lays down the hypothesis that slavery was the major cause of the faltering. Similar effects may be seen in the most advanced sections of the Southern States in the U.S.A. from 1825-1965 which were inhibited by slavery. Advance in the free North was rapid during the same period and accounted for its winning the Civil War.

Just prior to the British arrival in the eighteenth century, Sewai Jai Singh constructed a number of observatories and sent emissaries to all parts of the world before he embarked on the tasks of compiling *Zij-Muhammed Shehi*, his treatise on astronomy. One emissary brought back tables of Walter de la Hire, but the works of Copernicus, Galileo, and Newton were overlooked as was the invention of the telescope.[9]

The Madrasa at Calcutta was founded by Governor-General

Warren Hastings in 1781 and the Hindu College at Banaras in 1791, with little or no emphasis placed upon science. But by 1882 there were three institutions for medicine and three for engineering in India. Out of a total of 16,000 students in 1892, there were 1,950 students studying law, 800 medicine, and 500 engineering. The first science department at Baroda College of the University of Bombay was established in 1881, while the first science department at Calcutta University was begun in 1915. Elsewhere there was a school for surveying, established in Madras in 1793, and a Trigonometrical Survey of the Peninsula of India begun in 1800. In 1788 Botanical Gardens were founded in Calcutta, followed by the Botanical Survey of India the next year. Despite these salutary beginnings, it is only necessary to compare the development of Japanese science during the same period to realize graphically how little science was being pursued in an advanced civilization prevented by feudal and imperialistic economic circumstances from coming out of its medieval sloth or prevented by the indifference and outright opposition of the British.

By comparison the Meiji government of 1868 controlled the unprecedented (before socialism) development of Japanese science and technology since there was so little capital in private hands. This did not prevent the *Zaibatsu* (cartel-leaning corporations) from being encouraged, but that is another story. The Japanese experience in the development of science is markedly different from the Indian for such reasons as that Japan's ruling-class faced the challenge of Western technology as if it were a matter of life and death. Once it had decided upon a plan, Japan never quit until it had mastered science and technology, to be brought about by enormous expenditures on education, until then unique in history.[10] The diffuseness of India's ruling-class was a major problem for Indian development, even if that class had agreed that science and technology were of the greatest immediate importance. The spiritualistic soothsaying of the Hindu and Muslim revivalists could have been ignored, but was not, to the constant detriment of Indian scientific development. Then, as today, the ruling-class finds it convenient, as E.S. Namboodiripad has documented (and as he himself has to copy) to make political alliance with reaction and obscurantism even if the country should suffer thereby. Nevertheless,

Japan and India did have some common characteristics during the first period of modern scientific development, including population explosion, resistance to agricultural rationalization, slowness of the worker to effectively adopt technological inventions such as power, resistance to standardization required by capitalist production, and inadequate warehousing facilities[11]. I am not recommending that the worker should have reacted otherwise. In the light of what he knew, his feet-dragging reaction was perfectly rational. In comparing the Indian with the Japanese, the Japanese proved superior in centralization, superior in combining large with small scale production, controlling raw materials coming in from abroad, compactness of land communication, uniformity of cultural patterns, the existence of one language. India had the disadvantage of British Rule, but the advantages of a large railway system, a large body of skilled workmen, and above all, enormous natural resources.

One of the first notable Indian scientists before independence was Haffkine, a student of Pasteur, at the end of the nineteenth century, who developed plague vaccine and established the laboratory in Bombay which later developed into the Institute which today bears his name. Agriculture research began in 1905 in Bihar, but industrial research was not established until the first World War.

Founded in Calcutta in 1784, the first learned society devoted largely to science was the Royal Asiatic Society. The first scientific journal was *Asiatic Researches* (1799) begun by the Asiatic Society of Bengal also located in Calcutta. By 1900 there were thirty-two such journals which increased to 103 by the end of 1940.

INDIAN SCIENCE AND EXTENSION OF BRITISH SCIENCE

According to Abdur Rahman, Indian science "was primarily an extension of British science....and was at first limited to British personnel, but slowly and gradually the number of Indian participants increased and assumed responsibility both in departments as well as in societies and their publications."[12] As Indians participated more and more in the scientific enterprise, they created pressure to extend scientific education and research, national needs, requirements and aspirations. They also began to play a greater role in the evolution of policies and the growth of science. Rahman notes that the peculiar effects of British-introduced-and-controlled science continued even after Independence. These effects were : *first*, that science was not commensurate with the size and population of the country; *second*, since the teaching was in English, that is was limited to a small nucleus of people while the vast majority of Indians still were immersed in traditional medieval learning; *third*, that science teaching facilities were extremely limited, so that the separation of hand and brain occurred [and continued] because science was seen merely as another academic discipline; *fourth*, that the link of science with industry was weak because of the foregoing, and consequently, the government policy was itself based on conflicting interests and sometimes seemed scarcely aware of the needs of a scientifically based Indian Independence; and *finally*, that because of the paucity of facilities and trained scientific cadre, foreign experts, consultants, know-how, and equipment had to be imported. In a letter to me dated 5 June 1980, Rahman kindly pointed out to me that there were other complications : These were (1) that scientists in India accepted the prevailing capitalist philosphy of science without understanding its social implications; (2) that they limited themselves to the study of individual branches of science; and (3) that they were preoccupied with institution building. The need for specialists, preoccupation with nineteenth century conceptions of science, and infighting among scientists aggravated by the two-nation theory based on religion all contributed to the difficulties. No genius is required to notice that in such a complicated and ambiguous development, little interest would be generated in the philosophy or history of science—or in the science of science or the sociology of science. Indeed, these latter were scarcely taken up seriously until the end of the second World War by the most scientifically advanced countries. Nevertheless, Abdur Rahman initiated a *Science of Science Journal* in the 1960's, which he publishes from New Delhi.

ORIGIN OF THE PHILOSOPHY OF SCIENCE

Interest in the philosophy and history of science was generated by scholars concerned with both science and empirical philosophy. Several noteworthy, and by now classical exceptions to this rule were W. Whewell's *Philosophy of the Inductive Sciences* (1840), the *Dialectics of Nature*, written by Engels in 1895, but made available for English readers only in 1940 in the translation by Clemens Dutt, and V.I. Lenin's *Materialism and Empirio-Criticism* published in 1920. Appearing the same year was the forerunner of all later work in the philosophy of biology, namely the *Introduction to the Study of Experimental Medicine* by Claude Bernard. From that time onwards, there were fugitive attempts to lay out the broad framework of the philosophy of science, yet they were dominated by a passion for analyzing the methods of the physical or natural sciences. Consequently the history and philosophy of physics and astronomy in relation to methodology have been seen as having over-riding importance.

Just when philosophy of science begins in the West can be answered only within the framework of what one means by philosophy of science. J.J. Kockelman begins his work on philosophy of science by discussing Kant's contributions.[13] For Marx Wartofsky, philosophy of science seems to have appeared in Greek times since he treats the subject as a history of numerous problems such as laws, theories, induction, causality, explanation, and so on.[14] For J.D. Bernal, philosophy of science begins with Frederick Engels.

BEGINNINGS OF THE PHILOSOPHY OF SCIENCE IN INDIA

It was not until the second World War, that the totality of inter-relationships among the sciences was pointed out in a sophisticated way by J.D. Bernal in his epochmaking *The Social Function of Science* (1939), to be followed by his even more important *Science in History* (1954), second edition (1956), third edition (1968), since the latter work includes the discoveries and generalizations of the former. The Unity of Science movement begun by the logical positivists and logical empiricists was another contribution to awareness of philosophy of science and some of its special problems. However, in keeping with the general outlook of the contemporary positivists, the Unity of Science was presented in a piecemeal fashion in a series of small disconnected monographs by Bohr, Dewey, Russell, Carnap, and Charles Morris, Vol. I being *Encyclopedia and Unified Science*, ed. by Neurath. Absent were Infeld, Einstein, Bohm, Bernal, Needham, Hogben and J.B.S. Haldane.

The earliest efforts to inquire into, and give a cogent account of, scientific development in India, like those of B.N. Seal[15] and P.C. Ray,[16] were not as broad in scope as one would like. Many if not most of the studies since their work, have a bias in favor of science divorced from technology which has also been the trend in the West. There is direction in Indian science that is almost meditational and revelational according to Rahman, by which he implies activity of the mind divorced from activity of the hands. Encouragement to do experimental work required adequate facilities which the Indians surely lacked. The inter-relationship between the scientific and technological tradition in India has not only been studied, but its wider effects were not even considered until a few years ago. Ultimately, the great tradition of Indian technology and crafts will have to be incorporated into the history and philosophy of science. One could begin by including the material in G.M. Birdwood's *The Industrial Arts of India* (London, 1880), in two volumes and in Coomaraswamy's *Domestic Handicraft and Culture* (1910) and his *The Arts and Crafts of India and Ceylon* (1913).

SOME DIFFICULTIES IN DEVELOPING SCIENCE AND PHILOSOPHY OF SCIENCE IN INDIA

In the academic world it was Maurice Gwyer who first introduced history and philosophy of science at Delhi University in 1949-50. R.S. Gill, then at Panjab University Camp College, lectured on history and philosophy of science in 1954. He then taught the philosophy of science or perhaps, more correctly, he introduced philosophy of science at Delhi University at the invitation of Vice Chancellor C.D. Deshmukh. Gill, who had retired from teaching was sought out again in 1961 to begin the paper in Philosophy of Science at Delhi University, the same year that Panjab University Camp College abandoned the course. According to Gill, the major reason why philosophy of science has scarcely been introduced into the philosophy curriculum (or any other) is that traditionalism has such a tremendous grip on the Indian Universities which enables them, as he puts it : "to march forward looking back."[17] He told me that this traditionalism comes from the philosophy of gītā which was "pounded and washed into us," through the principle of *nishkam* or disinterestedness. "How can anyone achieve anything by being disinterested ?" Gill asked me in New Delhi in 1967. Gill's impatience perhaps came as a result of having been a railroad construction engineer in Burma. He added : "We are told to close our eyes to results, a stupid rule if there ever was one." Indian philosophy students are filled with *nishkam* according to Gill. What is needed he said :

> ..not only among its philosophers, is the sense of ratio of output to input, and attempt to maximize efficiency. In order to get that we cannot keep our eyes shut to results as the gītā recommends.[18]

In the meantime Abdur Rahman voices his complaint that even teachers of science are usually pencil-and-paper scientists, but this is sometimes true in the West as well. The actual preparation of some of these "scientists" is close to being merely literary. Behind this lackadaisical approach to philosophy of science and science itself stands Indian society and particularly the Indian Government controlled by the Congress and other essentially conservative parties. It is the view of Rahman, N. Sen, and N.R. Rajagopal that :

> First, Indian expenditure on scientific research in terms of percentage of national income, of government expenditure and of the capital investment is negligible. It does not even compare well with the small countries of Europe with a population of a few million. In terms of the total requirements of the country, as laid down by the [stated] objectives of the various plans, the investment in scientific research is also inadequate.[19]

In explaining the context of this situation, Rahman has pointed out to me in the same letter of 5 June 1980 that during the last year of Nehru's premiership, Indian industrialists had opposed the establishment of the Council of Scientific and Industrial Research as they thought that it would interfere with the free import of technology (see Battelle Report on C.S.I.R.). Meanwhile, their linkages had started attacking C.S.I.R. for not producing results and indicating that investment in research was wasted. Here again, the main purpose was to mount pressure for "free" and "unhindered" import of technology. If this followed the pattern of most Western industraialists, the aim was the quickest possible returns with no consideration given to the total movement of science and technology to help develop India. It is this syndrome in the West that has led to such heavy investments in armaments in which there is the greatest profit with the most outrageous pricing making oil profits appear moderate by comparison.

Responding to the unfair charge of the industrialists and their spokesmen, defenders of the C.S.I.R. pointed out that since investment for research was almost negligible, miraculous results could not be expected. It was hoped that the Indian Government would invest more in scientific research, provide better facilities for work, and have a greater control on the import of technology to avoid indiscriminate imports (some of which would, presumably, be counterproductive and gimmicky). Instead, every opportunity should be given to indigenous research. The piecemeal views of Hardy and other European scientists concerning the mythology of merely idealistic science had to be abandoned. Insisting upon scientific research as an instrument of development and social change was what was most needed. Searching questions must be asked about who is going to benefit from scientific research, the long-term goals of Indian society, and the seminal role of such research in the forward march of the whole Indian society with a de-emphasis upon immediate profitability. In every developing country there is the illusion that any and all technological importations will work miracles. The truth is that many of them are as useless as tractors of the American Great Plains on narrow hillside cultivation. Given encouragement, Indian technologists will make improvements and inventions that are suited to Indian soil. Although such a science policy can be recommended from the point of view of the practices of certain other capitalist countries which spend from sixty to ninety percent of their national budgets on military-related matters and hardware[20], the indolence in non-military affairs is disheartening, particularly in scientific agriculture which requires not only an agronomical solution but a political one as well. Bernal said some years ago and repeated recently before his death,

> The growth of science in the [advanced] capitalist world in the

> last few years has been phenomenal, but it has been at the expense of various distortions of aim and method.[21]

This would be true even if we credit military research for biological, aeronautical and space spinoffs, such as radar. But Rahman et al continues :

> Second, in the allocation as it is now being made to institutions, the major emphasis is on the institutional requirements rather than the requirements of different fields of specialization of scientific research. Consequently a picture of investment in different fields of science in relation to the national requirements and the growth of science is not possible.[22]

National requirements cannot be given priority because this would imply that the scientists themselves, instead of corporations and their politicians, decided what the national aims were to be. Under capitalism, the national aims are ultimately set by the owners of large corporations.[23] The Achilles heel of all the Indian five-year plans have been (supposed) corporate interests instead of national requirements. It should be remembered that despite the tremendous material gains of Japan from 1868 to 1931, corporation adventurism led by the *Zaibatsu* resulted in Japan's defeat and (temporary) collapse in 1946. Since that time Japan has become, ironically, the chief trade competitor of the U.S.A. which had encouraged her growth as deterrent to socialist advance in East Asia. What India's international role is to be in relation to China, the U.S.S.R. and U.S.A., not to speak of the Common Market and the Warsaw Pact Powers, is quite unclear.

Having stated their second point, we now turn to the third made by Rahman et al :

> Thirdly, more provision has to be made for capital investment in research to meet the requirements of equipment, library and other facilities than has hitherto been made, particularly in view of the heavy increase in cost of the new and sophisticated equipment and the need for modernizing facilities.[24]

It is difficult to know whether the relative indifference to science is calculated by the Congress Party leaders and its conservative advisers. Perhaps it is a shared indifference as a result of the prevailing social ideology and that it cannot, except for sudden conversion or a great enlightenment, be irradicated. The idealistic struggle in Indian philosophy, and to a lesser extent in Indian social science, religion and education (with few exceptions), appears to insure more than sufficient irrationalism. An outstanding example of this is to be seen in the writings and teachings of Sri Aurobindo whose influence among intellectuals who should be more critical has been and still is uncommonly great. We turn briefly to this topic.

II

RECENT PHILOSOPHIES OF SCIENCE IN INDIA

The most graphic and understandable introduction to the various philosophies of science to be found in India today is to give a thumbnail sketch of the direction and claims of the philosophers working in this field. But before we turn to them, it will be helpful for the reader to gain an insight into the working of a notable Indian mind who is well known in the West. Sri Aurobindo's followers have claimed that his outlook is particularly inclusive because, not only does he know religion, philosophy and the humanities, but also has a firm grasp on the sciences and their principles. From my first introduction to Aurobindo Ghosh disciples in South India in 1951, I was led to believe in the universality of his knowledge and understanding. At the same time I heard the important stories about him and the Mother who took over his Ashram when he died about that time. I was very impressed by what I heard. Later I met his partisans in the United States, mostly in Southern California. They had gained tremendous insights from studying his work they told me. They told me with joy that the daughter of Woodrow Wilson had been converted to the faith of Integral Advaitism. As a consequence I read about the Absolute as Consciousness-Force. Sri Aurobindo's writing reminded me of earlier readings I had done in New Thought, well-explained by Hundson and Troward in the United States.

What I liked about Sri Aurobindo was his recognition that there could be contradiction between consistency and truth, (between formal and dialectical logic). His solution to the contradiction (if there can be one) was that contradiction may be removed by the logic of the infinite. With the assistance of this logic what is the outcome in Sri Aurobindo for science?

Among Aurobindo's admirers there are some who seldom grow tired of speaking about his strong "scientific background and interests." On examination, however, we find that nothing could be further from the truth. His work demonstrates a consistent litany of scorn, if not palpable hatred, for the scientific enterprise.

SRI AUROBINDO AS FIGHTER AGAINST SCIENCE

That he is a revivalist, fideist, and idealist no one can doubt who has read his popular books. What he says in his *Letters*, however, is less well-known and indicates in a more intimate way his attitude towards science and his hatred of anything that smacks of naturalism or materialism. There are so many quotations that one could cite as evidence of this that the sheer wealth of material is an embarrassment. I shall try to choose only cardinal examples because, otherwise, this section would become unendurably long.

He says in a letter written November 18, 1934, at the height of his spiritual development:

> ..spiritual seeking has its own accumulated knowledge which does not depend in the least on the theories or discoveries of science in the purely physical sphere....[My] attempt like that of Jeans and others is a reaction against the illegitimate attempts of some scientific minds in the nineteenth century...who took advantage of the march of scientific discovery to discredit or abolish as far as possible the religious spirit and to discredit metaphysics as cloudy verbiage, exalting science as the only clue to the truth of the universe.[1]

In another letter he says:

> ..I think that attitude is now dead or moribund; scientists recognize..the limits of their sphere. I may observe that the conflict between religion and science never arose in India (until the days of European education) because religion did not interfere with scientific discovery and scientists did not question religious or spiritual truth because the two things were kept on separate but not opposing lines.[2]

This alleged lack of conflict between religion and science or between theology and science, as the cautious A.D. White used to point out, does not square with what M.N. Roy, himself no paragon of scientific thought, points out regarding science and Hinduism. Says Roy:

> To have found the unity in diversity, is claimed as the greatest merit of Hindu philosophy. But, as a matter of fact, the unity was not found. It was simply assumed or imagined. It is an ideal conception which brushes aside the problems to be solved. Since the rise of the material world out of the assumed immaterial root-cause is not logically possible, dualism persists defying all metaphysical verbal jugglery.[3]

Roy fails to note the verbal jugglery perpetrated by dualism. Where once it served a progressive function for Descartes, today it leads only to mischief. R.G. Collingwood at Oxford tried to demonstrate that both religion and science have their respective spheres of influence, a view also claimed by Sri Aurobindo. Religion, he said, deals with the infinite and science with the finite. Therefore there can be no conflict. Collingwood's next step is to equate the infinite with the Christian spiritual apparatus in his *Faith and Reason* (1948). Aurobindo does not use this argument. But he does take Jeans and Eddington at face value as we shall see P.J. Chaudhury and Ruth Reyna also doing when they write on the philosophy of science. Jeans and Eddington revealed the "new" spirit resulting from the so-called "spiritualization of matter," where matter, by some mysterious and inexplicable transmutation, becomes mind, consciousness, or spirit. They were not alone in this. Spiritualists sprang up in Europe, the United States, and Africa (followers of General Jan Smuts) who also found solace in these views which were subsequently refuted by L. Susan Stebbing in her *Philosophy and the Physicists* (1937). Aurobindo's reading in anti-spiritualist, critical, literature was at most perfunctory while he was still a young political "radical". Even today in India, it is so difficult to find scientifically critical works except in the relatively scarce radical bookshops that one could not imagine Aurobindo frequenting after his conversion to all-out spiritualism. His knowledge of the struggle between science and religion was at best miniscule, especially as it referred to the Indian experience from Lokāyata onwards. Outstanding among Aurobindo's spiritual qualities are his avoidance of the history and development of science—or civilization for that matter. He says, for example, that there are periods in which there is a temporary rise of materialism. Such periods:

> ..came in Europe in the nineteenth century [and the 18th, the 17th, the 16th, etc.], but they are usually of short duration. Western Europe has already lost its faith in materialism and is seeking something else [not twenty years after this pronouncement, 37 percent of the people of the globe were officially or unofficially allied to materialist philosophy], either turning back to old religions or grouping for something new. Russia and Asia are now going through the same materialistic human development—to destroy the bondage of old forms and leave a field for new truth

> and new forms of truth and actions in life as well as for what is behind life.[4]

This is based upon at best a shallow view of contemporary history. It makes us aware of the lack of success revivalism has when it is faced by knowledge instead of ignorance. When Aurobindo wrote this letter in 1935, the amount of materialism in Europe had doubled and China was just about to carry its masses into the materialist orbit. He does recognize, nevertheless, at this time, that there is a "certain necessity in human development," in him a relatively healthy sign of historical idealism that echoes what Hegel said before 1830.

Like most religious devotees Aurobindo is concerned about free will versus determinism. In pursuing the problem of predeterminism, he says:

> ..behind visible events in the world, there is always a mass of invisible forces at work unknown to the outward minds of men [men have inward and outward minds?], and by Yoga (by going inward and establishing a conscious connection with the Cosmic Self and Force and forces), one can become conscious of these forces, intervene consciously in the play, and to some extent at least determine things in the result of the play.[5]

Since invisible forces are generally not known, being invisible, we are left with visible forces, which is precisely what scientific method points out. But, says Aurobindo:

> All that has nothing to do with predetermination. On the contrary, one watches how things develop and gives all that to contradict a dictum of the great scientist [scientists are great, when they can be quoted for spiritual propaganda], C.V. Raman. Raman said once that all these scientific discoveries are only games of chance. Only when he says that scientific discoveries are games of chance, he is merely saying that human beings don't know how it works out. It is not rigid determinism, but it is not blind inconsistent Chance either.[6]

It is pitiable that Aurobindo has to paint C.V. Raman with his obscurantist tar brush, but if Raman has fallen into the positivist pit (fashionable enough in his day among natural scientists) by denying that we know a real world in science, then Raman must take the consequences for this lapse just as Eddington and Jeans must. The last sentence of Aurobindo, concerning "Chance," gets us into the realm of rubberization of determinism. Instead of rigid, inconsistent determinism, we have flexible, malleable, tractile determinism, not simply "blind inconsistent Chance.." Yet, what does this imply other than that chains of determinism can be altered by other such chains? We are quite satisfied in this context to have Sri Aurobindo admit that it is not mere chance.

Concerning "The Revolution in Science," which is the heading of one of his letters, Aurobindo shows his increasing resourcefulness in equipping his revivalist arsenal:

> The defect in what X writes about Science seems to be that he is insisting vehemently on the idea that Science is still materialistic or at least that scientists, Jeans and Eddington accepted, are still fundamentally materialistic. This is not the fact. Most continental scientists have now renounced the idea that Science can explain the fundamentals of existence.[7]

Now, let's see where we are. Most scientists, it is alleged by X, still believe in materialism. Instead of proving this to be false by a scattering of citations, Aurobindo reduces it to ashes in the fashion we expect of philosophers who are in touch with the Cosmic Self. He quickly shifts ground, having left that notion dangling and turns to the idea that most continental scientists [Aurobindo would have surprised many of us if he had known ten of them by name, when there are thousands to interview on this question], renounced the idea that science can explain the fundamentals of existence. If we take the fundamentals to be matter in motion, then it is up to Aurobindo to prove that most scientists remounce them. It is true that under the subjectivist and solipsistic influence of positivism, quite a number of scientists who wrote about it had doubts as to whether they were talking about ontological conditions or statistical events [for them statistics being *of nothing* or *mere concepts* since they were not about material reality]. Whether even one percent or ninety percent had grave doubts we shall never know, since no surveys were made at the time. As Leopold Infeld expressed it in *Quest*, his autobiography, "Einstein and I believe every scientist's approach is that of a realist. There is nature, there are laws, and the scientist with his feeble mind tries to form a consistent, logically simple picture of reality, a picture which can explain the world of our sensual [sic] impressions."

Let us continue with the same letter to get the full flavor of Aurobindo's knowledge of science:

> They hold that Science is only concerned with process and not with fundamentals. [But process is fundamental, as even the early Buddhists and Heraclitus intuited]. They declare that it is not the business of Science nor is it within its means to decide anything about the great questions which concern philosophy and religion. This is the enormous change which the latest development of Science has brought about. Science itself nowadays is neither materialistic nor idealistic. The rock on which materialism was built [materialism *is the rock upon* which everything else was built] and which in the nineteenth century seemed unshakeable has now been shattered. [Where has it been shattered and by whom?] Materialism has now become a philosophical speculation just like any other theory [this view of theory is like Newton's uncomplimentary view of hypothesis—an unfounded metaphysical guess. Scientific theory is based on a truthful account of reality]; it cannot claim to found itself on a sort of infallible Biblical authority, based on the facts and conclusions of Science.[8]

What *is* Aurobindo talking about? He is to science what President Reagan is to Soviet-American diplomacy. Certainly he is not writing about twentieth century science, but about speculation based on the vedas. And we can hardly object to his knowing so little about science, recognizing what a difficult time Indians have had to introduce it. But we can object to his writing as if he were conversant with it when what he knows is worse than knowing nothing, since it is all wrong. To continue with his soliloquy:

> This change can be felt by one like myself who grew up in the hey-day of absolute rule of scientific materialism in the 19th century. [This certainly does not square with the history of the reception and spread of scientific views in the Bengal where Aurobindo grew up]. The way which had been almost entirely barred except by rebellion now lies wide open to spiritual truths, spiritual ideas [As if Bengal, the most enlightened part of India, was not choking in spirituality!].[9]

It is interesting to note in passing that as Aurobindo was writing this letter, the Nazis in Germany were putting into practice the spiritual philosophy of Alfred Rosenberg—see his *Der Mythus Des* 20 *Jahrhunderts* (1935). Beyond the simplistic and narrow world of pathetic materialism, important events are stirring, Aurobindo tells us. He says:

> That is the real revolution. Mentalism is only a half-way house but mentalism and vitalism are now perfectly possible as hypo-

> these *based on the facts of existence* [my italics—DR], scientific facts as well as any others. [Are there non-scientific facts, marching along without their uniforms?] *The facts of Science do not compel one to take any particular philosophical direction* [my italics—DR]. They are now neutral and can even be used on one side or another though most scientists do not consider such a use admissible.[10]

Well, who cares what most scientists think, as long as we are in touch with the Cosmic Self. Tyrants in history begin to sound modest next to the arrogant pronouncements of Sri Aurobindo, super-scientist. He has now contributed to science the doctrine of double-facts to add to the doctrine of double-truth, the doctrine of double-self, the doctrine of double-reality, and the doctrine of double-talk.

At this Juncture, as useful as Sir James Jeans[11] has been to Sri Aurobindo and spiritualists throughout the world, Jeans draws a sharp counter-attack from Aurobindo when the British physicist suggests that life came into existence on earth by accident. Aurobindo warns us about such statements as "these are mere mental speculations [as opposed to physical speculations?] without any [material ?] conclusiveness about them...." Opposed to such mere surmise are *the facts* [scientific or non-scientific or unscientific] put forward by Aurobindo to wit:

> The material universe is only the facade of an immense building [larger than the Pentagon or the Merchandise Mart] which has other structures behind it [perhaps like the Rashtrapati Bhavan in relation to the Lok Sabha] and it is only if one knows the whole [like Sri Aurobindo] that one can have some knowledge of the truth of the material universe. There are vital, mental and spiritual ranges [like the Himalayas, Western or Eastern Ghats] behind which give the material its significance. If the earth is the only field of the spiritual evolution in matter [what?]..then it must be as a part of the total design. The ideas that all the rest must be waste is a human idea [we hoped that it was not a divine idea] which would not trouble the vast Cosmic Spirit.

Aurobindo can know this only because He Himself is the Vast Cosmic Spirit, for otherwise who can know what would trouble Him? As one moves from stage to stage, from range to range in that Great Himalaya in the Sky (in spiritual evolution), one gradually comes to "greater Consciousness" of a cosmic existence in which form, life, and mind no longer appear by accident [mind seems to be accidental in the coming into being of Aurobindo], but "find their significance." The two stages in which one is involved are first the "overmental" and second the "supermental." Only at the second stage can one understand the last full truth of existence and become truly aware of its significance, says Sri Aurobindo.[12] Considering the cosmic intelligence required to know all this, it is astonishing that Aurobindo could not even predict what the Mother wanted to have for lunch.

I wish now to quote some arresting statements from Aurobindo's illuminated writing to give the reader an even clearer idea of his outlook on the scientific method and enterprise:

(1) About spiritualism, I think I can say this much for the present. It is quite possible for the dead or rather the departed—for they are not dead—who are still in regions near the earth to have communication with the living.[13]

(2) The case of astrology is fairly strong; a case seems to exist for Cheiromancy [palmistry] also.[14]

(3) About polytheism, I certainly accept the truth of the many forms and personalities of the One which since the Vedic times have been the spiritual essence of Indian polytheism.[15]

(4) Mother India is not a piece of earth; she is a Power, a God-head, for all nations have such a Devil supporting their separate existence and keeping it in being.[16]

(5) Psychisation means the change of the lower nature bringing right vision into the mind, right impulse and feeling into the vital, right movement and habit into the physical—all turned towards the Divine, all based on love, adoration, bhakti—finally the vision and sense of the Mother everywhere in all as well as in the heart.[17]

(6) The Jivatman, spark-soul and psychic being are three different forms of the same reality and they must not be mixed up together, as that confuses the clearness of inner experience. Comment: As outer experience gets more unclear through the dismissal of science, the determination to get clear inner experience increases.[18]

(7) If the psychic were liberated, free to act in its own way, there would not be all this stumbling in ignorance. But the psychic is covered by the ignorant mind.[19]

(8) [Man] carries about with him an environmental consciousness [called by the Theosophists the Aura] into which they

first enter.[20]

(9) It is only if one enters into the Cosmic Consciousness that one begins to see the forces at work and the lines on which they work and get a glimpse of the cosmic Self and the Cosmic Mind and Will.[21]

(10) The spiritual mind is a mind which, in its fulness, is aware of the Self reflecting the Divine..when the mind is turned towards the Divine and Truth and feels and responds to that only or mainly, it can be called psychic Mind.[22]

As in the case of adepts, Aurobindo has synthesized his own private "science" including private psychology, private cosmology, private metaphysics, private anthropology, indeed in any "science" where he finds confirmation for his ideas. Whereas psychoanalysis seems to explain by pulling outside forces inside, Aurobindo appears to pull the inward forces outside toward the supermental and divine. This special proclivity is associated with revivalism of a solipsistic kind as A. Rahman has noted in his *Science, Technology and Economic Development*.

Sri Aurobindo's treatment of sex is of particular interest because of his knowledge of Tantra and hence his awareness of its importance in the development of Indian ritual and thought. He says:

> The sex-impulse is certainly the greatest force in the vital plane.. if it can be sublimated and turned upwards, *ojas* is created which is a great help to the attainment of higher consciousness..sex.. is a movement of general Nature seeking for its play.[23]

If sex is entirely controlled, says Aurobindo, then it can be turned into spiritual energy force—force even beyond productivity and artistic activity. Thrill in sex is "simply a gross distortion and degradation of the physical Ananda which by the Yoga can establish itself in the body.[24]" For supermental activity, one renounced indulgence in sex, although it is not so easy to stop having sex dreams, for, as Aurobindo admits:

> I have heard it said that even very advanced Yogis get the dreams at least once in six months—I don't know how far it is true or what the Yogis themselves know about it.[49] But the sex-impressions in the heart can be got rid of long before the end of life [that

> *is* the end of life], and even the seed state in the subconscient which comes up in dreams, though sticky enough, is not quite irremovable as all that.[25]

If there are no dreams, discharge should not be fretted over because it may have "purely material causes, e.g., the pressure of undischarged urine or faecal matter near the bladder."[26] This is an unexpected admission, for it indicates that Aurobindo is not yet a total idealist, regarding urine and faecal matter as *māyā*. Nevertheless, he admonishes us, the main thing under such circumstances is to remain undisturbed. If he can remain undisturbed under such circumstances, he may as well remain undisturbed while having purely material venery. If one can gently pressure by will or force the physical subconscient in the dream-emission, then one can do it in intercourse—especially if he has the co-operation of a highly conscient and spiritual woman nearing union with the Cosmic Self. However, Aurobindo warns, if the subconscient is obstinate, then one must simply keep applying more pressure on it. Bit by bit sex-thoughts must be extruded and rejected by the mind by the higher vital, and finally by the environmental consciousness which is a "surrounding atmosphere which we carry about with us and by which we communicate with the universal forces" until sex-thoughts just dry up and blow away. A final warning to the leisure-class *brahmin* and *kshatriya*: "Inactivity is an atmosphere in which sex easily rises.[27]" And of course this is why full-time spiritualists keep thinking about it, being occupied with unmateriality. Fortunately, Buddhists such as Chogyam Trungpa regarded spirituality as a way of making ourselves comfortable, hence spirituality must be shunned according to him.[28]

Sri Aurobindo's fascinating view of science is certainly pre-Thalian or pre-Carakan. He has chosen to remain almost miraculously ignorant of the whole subject and its history, not to speak of its method and its hopes for the future.[29] His reaction, however, is generally no more striking than that of other Indian revivalists whose words are taken with deadly seriousness by many literate people.[30] One would be seriously mistaken if he thought that Indian professors of philosophy were unaffected by his idealistic and anti-scientific outlook.[31] With such opinions having wide circulation it is easy to see that Indian science and philosophy of science still face some challenging obstacles.

PHILOSOPHY OF SCIENCE IN INDIA TODAY

For this section, I have chosen certain representative figures who recently have contributed to studies in the philosophy of science. Such work is so uncommon that one does not have to pick and choose with great diffidence.

PRAVAS JIVAN CHAUDHURY AND TRANSCENDENTAL SOLIPSISM

Few Indian institutions of higher learning offer course work or papers in the Philosophy of Science. Since there is so little demand for textbooks and supplementary reading material, publication in this area is slight. One of the few books to appear under the rubric of philosophy of science is Pravas Jivan Chaudhury's *The Philosophy of Science*.[32] As is generally the case, Chaudhury's work is more concerned with physics than with science as a whole. This has meant, both in India and in the West, that philosophy of science is given a peculiar and distorted perspective, leaving out the chemical, biological, and social perspectives. One might speculate as to the reasons for this. Part of the reason is that physics is more accessible to the layman in popular works which require no laboratory work. It is a relatively "clean" science ever since mathematics became so enormously important to it. Chaudhury is not concerned with science as a whole, rarely mentioning that there are other areas of human interest besides physics and astronomy. Consequently he never raises any interesting questions about biology and social science. In the Introduction to his book, Chaudhury gives us a clue as to the kind of philosophy of science that he is likely to espouse. He here says:

> ..though the philosophy of science (including the history of methods of science) has been recognized to be a prominent branch of philosophy in the West, in India, the subject has not been taken up yet, owing perhaps to the predominantly metaphysical or esoteric bent of the oriental mind.

He gives us another clue to his presuppositions by stating that".. the present author holds..neither [a view] of radical positivism nor one of dogmatic idealism but one which may be termed rational empiricism."[33] On the same page he says that the function of the philosophy of science "....is to judge the very bases of science and a constructive scientific philosophy, and so, adopt the critical and intuitive method."[34] Philosophy of science, he continues, also means "..inductive metaphysics..philosophy of nature and not..spirit or ..total experience."[35] Again, he says that "..philosophy of science [is] a critical reflection on science, and so, following a rather intuitive method instead of the inductive one [hence this] may give us more comprehensive philosophy."[36] Chaudhury maintains that the difference between deductive metaphysics, which must be rejected by philosophy of science, and inductive metaphysics, is that the latter is "self-corrective and open to permanent control by facts."[37] We hope that Chaudhury is next going to tell us how we can determine what are the facts. His answer is not long in coming, but it is not meant to be a direct answer to this question. "Matter," says Chaudhury, "is an order of abstraction for which modern physics finds no adequate operational rules for verifying..and so, has banned it from the vocabulary of physics."[38] Ever since 1710, when Berkeley rushed *The Principles of Human Knowledge* into print to be in time to begin modern obscurantism, matter has been banned by idealists. Evidently, Berkeley himself had a touch of that "metaphysical and esoteric bent."

At this point we must ask Chaudhury a question so simple that it is almost embarrassing. This question is: Just what is matter an abstraction *of*? If there is an abstraction, then it is an abstraction *of something*. Now concerning the head of the Cheshire Cat: "The King's argument was that anything that had a head could be beheaded. The executioner's argument was that you couldn't cut off a head unless there was a body to cut it off from...."[39] So, If you have an abstraction from something, then there must be a something that the abstraction is from. Now, if Chaudhury dared answer this query, he would have to rewrite his entire philosophy of science; if he didn't dare, then I find no imperative, including politeness, to continue to read his philosophy of science since it turns out to be simply a form of subjective idealism which has been skillfully dealt with by L. Susan Stebbing in *Philosophy and the Physicists* (1937), and by Lenin before her in *Materialism and Empirio-Criticism* (1909). Using the Principle of Least Effort, pared out by Occam's Razor, I see no reason for refuting this view again. One can only repeat that if there are sensations or abstractions, that they are sensations or abstractions *of something*. Bluntly put, matter is not an abstraction at all, but that base from which abstractions are drawn. It is Chaudhury's non-thinking thought of matter which is an abstraction from matter. Matter is by no means an abstraction of his thought of matter. Some reader will say that it is unfair to pick on Chaudhury because he is rather naive, but I respond that philosophers of science both East and West are naive when they think that they can dispense with matter (energy).

Chaudhury says that "science cannot be used to back up a materialistic thesis"[40] anymore than an idealist thesis. So, dropping between two stools, Chaudhury adopts the imaginary neutralism of positivism, since "imagination is the weapon against reality" as Jules de Gaultier pointed out. Furthermore, says Chaudhury, whose philosophy of science begins to look like an exercise in free-association, the new physics "cannot prove the physical universe to be non-mechanical and material,"[41] this, although he has just said that "Nature to science is a real objective entity, a solid reality to be faced, to be exploited and fought at once."[42] But Chaudhury has already assumed that it is material if it is physical (non-mental) unless we postulate a third category (along with Popper, Lakatos, Mādhyāmikha, Plato)—some neutrality, for which there is even less evidence than there is for the mental, as Aristotle demonstrated in his *Metaphysics*, 990-993. Despite all this Chaudhurian confusion, he then goes on to say that perhaps there is :

> ..some less determined stratum of matter..in the cortex and acting as a *via media* between the two spheres, the free mind and the physically determined nature.[43]

He has just plopped us into the middle of Plato's third man argument, pointed out as an indissoluble mare's nest by Aristotle. For if you have a P connected to Q by an R, then you have P connected to R, and Q connected to R, and that in its turn gives you a new connection to explain and so *ad infinitum*. To avoid this needless paradox, one might just as well accept pre-established harmony.

One may justifiably suspect that philosophers like Chaudhury, indeed, many idealistic philosophers, are doing their level best to make science so inscrutable that everyone will turn, out of desperation, to theology for succor. Concerning this, I have attended conferences at which research scientists listened to philosophers of science hold forth, scratch their heads and say that if they had known that science was so difficult that they would have gone into basketweaving. A major problem is that philosophy of science practitioners, both in India and the West, are reduced by their ineptitude, to talk about how painfully difficult if not impossible it is to do science, especially if, like simpletons, we are unable to accept a materialist view of the external world. The usual cry in the West is, after slogging around in the mire of positivism and neutralism sufficiently, "back to Kant! " That means that we admit that we can never know any material world, but are reduced to the contents of our consciousness.

Before dropping an examination of Chaudhury's confusions, I wish to set down his final word on philosophy of science which appears on the last two pages of his work, *The Philosophy of Science*. Says Chaudhury,

> We conclude our deliberations, then, with a statement that though science and such philosophy as bases itself on its results, e.g., inductive metaphysics, cannot itself prove any idealistic or spiritual philosophy yet it is the expression of an attitude which is not ultimate but relative to a more ultimate attitude which comes over when the former is critically reflected upon.[44]

I shall stop here so as to break up this tortuous paragraph. Beginning again,

> The latter attitude is idealistic and directly shows itself to be more fundamental than the realistic one which is an assumed self-delusive mode of the self or spirit in its love of variety and sport.

We remember that Kumārila had said that God could not have created the universe, since it has no purpose. Bādarāyana replied that God created the universe out of sheer *sport*. But Chaudhury continues :

> The critical reflection is itself a sign of the spirit's becoming weary of the sport [note that Chaudhury, like Sri Aurobindo, knows the "mind" of the spirit] and its gradual coming back to its own from the lapsed state. [Chaudhury has now become the spirit's psychiatrist.] We thus find that science has as its logical and psychological basis, a realistic faith, but it leads and gives way to an idealistic one through a process of transcendental criticism

> that naturally follows it (the latter's) object. In this sense we may say that science is not antagonistic to idealism, but leads to it as all relativity does to the absolute.[45]

I leave to the reader to explain this mumbo-jumbo. When we disown the material world, we are forced to populate the vacuum with an imaginary world of weary spirit, lapsing into the transcendental. As Juvenal put it, under such circumstances "it is difficult *not* to write satire." (*First Satire*, line 19).

RUTH REYNA AND THE STRENGTH OF NEGATIVE ATTRIBUTION

Another philosopher in India working spasmodically on the philosophy of science is Ruth Reyna, who published *The Philosophy of Matter in the Atomic Era*.[63] There is no doubt that her intention is to tie together traditional Indian philosophy and certain trends in modern science. She believes that the way to do this is to point out the idealism in both. She says :

> The idealistic urge in the scientific conception of the physical world is seen to arise out of the historical developments of science..[i]t is no small task to expound physical science in the frame of idealistic philosophy.[46]

Yes, indeed. One must agree that this is herculean. But she continues with admirable clarity :

> The total point of this work is to testify to the idealistic validity of the Vedantic concept of *maya* as an explanation of the relationship between appearance and reality, in which the phenomenal world is held to be neither real nor unreal and at the cosmic level is non-existent.[47]

First, a logical point, if the phenomenal world is nonexistent at the cosmic level then it is nonexistent at every sub-level of the cosmos. Next, one is led to believe that Reyna shows great restraint and modesty in her use of the terms "testify to" rather than "to prove" or "to demonstrate." But there is no reason for this modesty which begins to appear less so when she states that "neither man nor philosophy rests content with the materialistic or the empiricistic explanation of the world."[48] I agree with her that many of us are not content with the empiricistic non-explanation of the world, but must sharply disagree concerning her contention about the materialistic explanation.

There are thousands of scientists and dozens of philosophers of science (in the socialist world alone) who accept the materialistic view—which simply contradicts her false assumption. One can understand why she might mistakenly think this about the materialistic explanation when one turns to her bibliography which includes Eddington, Jeans, and an array of idealistic philosopher-physicists from our old friend Pravas Jivan Chaudhury to von Weizsäcker all members of the *Internationale Idealiste*. Reyna, in attempting to establish the thesis that the literature of contemporary science testified to the correctness of her idealistic thesis, gives us no references or substantiation that effect.

> The broadened [narrowed?] scope of particle physics and its resulting dematerialization of the atom has brought science to look upon phenomenal events in a double frame of negation—that the material world is neither real, nor, on the other hand, is it unreal. [Reyna wisely refrains from defining "reality"]. And although such negatory references cannot be found so literally stated in the writing of modern scientists, it can, beyond doubt [!] be inferred from their conclusions, as the present treatise fully demonstrates.[48]

Having herself postulated that matter is neither real nor unreal, Reyna then asks : just what is it ? No one can say, she answers. By way of apology, she adds that "It is quite possible to say the phenomenal world is neither real nor unreal without violating logic..."[49] Doesn't it depend upon how you define and interpret logic ? If logic has an ontological base in the material world which it must have to deal with reality, then to say that the phenomenal world is neither real nor unreal is to violate logic. If you agree with the American impalpabalist, Ernest Nagel, that there is a logic without ontology, you are enabled to make an infinite number of statements with no connection to the world and revel in your own phantom luminescences.

Reyna now lays down another principle that gives us one more clue to her methodology. She says, by way of making it more palatable that matter is neither real nor unreal, that "....if we say what a thing is not, we know something about it in a positive way." She gives no example of this surprising statement, so I shall supply one for her : Matter is not phlogiston. Now what is positive about that ? If we say what a thing *is* do we say something about it in a negative way ? That would seem to be parallel to maintaining that we say something positive when we say something negative. Like all subjective idealists, Reyna would like to crawl out of the designations discovered over three million years ago by mankind and retreat into the neutralistic "knowledge" of algae. Well, never mind, because she also says that:

> ..we have postulated that the world [things] has neither being nor non-being. It is, in other words, inexplicable. And that is just what science is presently saying.[50]

We still need to find some quotations from various scientists to prove that this *is* "what science is presently saying." She herself is dissatisfied with her own position, so a few sentences later states that the world is "an unexplicable [sic] 'force'." One wants to say that *her* "world" is an inexplicable 'farce'. But what a fascinating suggestion—that something inexplicable *has force.* If it is inexplicable we cannot predicate anything of it ! Really, Reyna is honor-bound to retract "inexplicable" and replace it by "almost inexplicable." Fair is fair. If we know one attribute of a thing then it is no longer inexplicable. Not satisfied with this logical, semantical, and ontological blunder, she then expands the inexplicable X as "an analytical apprehension." In other words, when I look at viruses through a electron microscope, the wiggly lines and shapes represent an "analytic apprehension," which is also what I get when I look through a telescope at the satellites of Jupiter. Since "apprehension" can mean either "perception" or "conception," I presume that she means "*perception*" of the almost inexplicable. So now the inexplicable is perceptible in addition to being a force. This inexplicable, perceptible force is, Reyna says next, "the substratum of reality." I must review the steps we have taken : the inexplicable, perceptible force serves as the substratum of reality ? Now, what could she mean by such a statement or postulate, which is self-contradictory ? Well, she must mean that the world or universe is a substratum of reality—a layer of some kind of material holding up reality. But that merely adds to her and our confusion. First, she says that we have an inexplicable substratum which implies of course that we cannot know that it is a substratum which in addition to being a SUBSTRATUM also has the properties of being PERCEPTIBLE, of being FORCE. On top of this inexplicable something we also have a reality something held up by a layer of inexplicable substratum. Reyna tries to allay our uneasiness or risibility by saying :

> We have but to turn to Buddhism, or to the Vedantic philosophers, especially of the scholastic era, to recognize that their explanation of the universe, lying latent for some twelve centuries [not lying latent, but rejected], is that which has been recaptured in modern science.[51]

What an amazing and wondrous situation—this recapturing by science of the hoary truths of [Mahāyāna, Mādhyamika or Yogacāra] Buddhism, and Vedānta. It just goes to show you ! What on earth has science been doing all those centuries ? According to Reyna, it has been stripping itself more and more of its unfortunate materialism, until it has discovered that reality is being held up by an immaterial, inexplicable force. As if to give us the strength and hope to go on, despite this devastating revelation, Reyna assures us that :

> In claiming that the world is neither real nor unreal, the sages are far less naive in their outlook than it seems to us who measure sophistication in terms of semantics.[52]

Speak for yourself, madame. I find that this hardly fits my own case. I measure sophistication in terms of whether someone can separate reality from non-reality, the sheep from the goats, someone who knows that there is an external, independently existing material world. If a person bases his thought on material reality he need not base it on the sophistication of sophistical semantics. Semantics, in any case, must ultimately rest upon ontology.

Reyna sallies next into the dark forest of the "Methodology of Negatory Metaphysics" with all the screeching denizens warning each other of her approach. She maintains that :

> ..a systematic use of negation that *absorbs realism* [my italics—DR—] and arrives at the residue of "no attribute" or *distinctionless-ness*. Negation in this sense is not negative in the meaning of "void", and is the only efficiacious method of affirmation. In defense of this broad statement, I have to remark, that I do not see why words of affirmation should be considered descriptive of a greater real than words of so-called negation [they are not so-called] that when pushed to their furtherest limit leave the real exhibited in its true character, intuitively known and unhampered by the limitations of the intellect.[53]

One must say to Reyna again : speak for yourself, speak for the limitations of your own intellect. Since I know some limitations of my own intellect and your Reyna-intellect, I depend upon science, which is objective and positive and not gibberish. You don't absorb realism by this method, you destroy realism. Then you find yourself wandering around in the great indistinct abyss of intuitions unable to describe anything. This is what happens to the intrepid person who shuns reality. To help Reyna get out of this *cul de sac*, the Chinese philosopher Chi-tsan (549-623) announced, says Reyna, the doctrine of double truth. There is truth in the common sense way and truth in the higher uncommon sense way. According to Reyna :

> The man or woman of meager development in scientific wisdom views the phenomenal as really *yu* presumably the realist, that is, as having being or is existent, and knows nothing about *wu* lucky fellow, or the non-existence of the world. . a higher sense of truth and a step forward in the progress toward a cosmic [comic?] realization.[54]

Evidently we are back with Sri Aurobindo, stepping into the cosmic. That we are always in the cosmic never seems to cross Reyna's mind which may be described as filled with *wu.* But she has more to tell us about *yu* and *wu.* It is one-sided, she says, to say that all things are *yu* or *wu* because *yu* is really *wu.*[55] This is the case because *yu* ceases to exist every moment, thus changing into *wu* based on the doctrine of mometariness, a basic postulate of Buddhism. Awareness of this state of affairs leads to the higher stage of the revelation that the world is neither *yu* or *wu* and can be talked about only in negative terms in which nothing can be affirmed.[56] If the dicta of this Buddhist School were followed by science, the laws of matter in motion would not be written down or enunciated since the higher truth is that we can affirm nothing. But, if we try to apply this higher truth to science in a concrete way, we would have to rewrite such definitions as found on the first page of Newton's *The Mathematical Principles of Natural Philosophy* as follows :

> Definition I. The quantity of matter is neither the measure of the same, arising from its density and bulk conjointly nor the absence of the same arising from its density and bulk conjointly nor the absence of the same arising from its density and bulk conjunctly.[57]

If Newton had written this, we would be disgusted. But let's take a more homely example:

A PEDAGOGIC DIALOGUE

(Ruth Reyna is at the market talking to the grocer)

Reyna : "How much are the mangoes ?"
Grocer : "They are not two rupees."
Reyna (impatiently) : "How much are they then ?"
Grocer : "Since I don't wish to appear to be semantically sophisticated, I'll say that they are not three rupees."
Reyna : "Good, this demonstrates that negation is just as efficacious as affirmation. By the way, how much are the mangoes, really ?"
Grocer : "They are not four rupees."

Clearly, negative attribution would be not only the death of science but of all meaningful communication. If Reyna wishes to pursue her Vedanta, her Ch'an, her Zen or whatever, no one can yet stop her. But there is no reason why she should intrude such irrationalism into the arena of philosophy of science. As L. Susan Stebbing has shown and Lenin pointed out earlier, there is not the slightest reason to believe that matter has vanished or that idealism is true. According to Stebbing, there are two causes for the belief that materialism is untrue. She says:

> Two main factors are responsible for the tendency of modern physicists to adopt some form of idealism. There is first the change in the conception of matter; secondly the change in the status assigned by many physicists to natural laws.[58]

There is a change *in the conception* of matter—more is known about it, but there is not the slightest evidence that matter or the physical has become any less substantial than it ever was. The very term "substance" and "substantial" depend upon this truth. According to the German physicist Karl Menger, "matter has an atomistic structure....or is at least composed of lumps."[59] In the Einsteinian equation, $E=mc^2$ which states that energy is equal to the mass times the velocity of light squared, it is assumed that the mass and energy are forms of matter. Matter has energy and matter has mass, otherwise there could be no *measurable* equivalence. Reyna would have measured the absence of things. There is a mass of *something* and an energy of *something* which is open to being measured by the senses and this *something* is matter. We cannot lose matter by putting it into an equation. Does it make sense to say that physical equations express nothing about anything ? That is what Reyna implies in her negative position. According to H.A. Lorentz (1853-1928), matter is the seat of electrical charges. If you ask, "What are those charges of ?" the answer is that they are charges of something called matter.

Physics is still the science of matter or energy in motion just as it always has been.[60] It is not the science of negative attribution *wu*. It is certainly not a non-science of the inexplicable non-substratum of non-reality. The belief in materialism has one of its roots in the belief based on experience of matter in motion that there is a universality of laws in accordance with which material bodies behave. Another root is that matter is indestructible—this knowledge is also based on mankind's long experience. Its operational meaning is codified in the first law of thermodynamics which no idealistic physicist has refuted.

In her "Apologia" at the end of her book which she should have entitled *The Impossibility of Science and the Disappearance of Matter*, Reyna states that "....the over-all governing postulates of physical science, strangely enough, can never be demonstrated or realized by finite mind."[61] What we must do to get ourselves out of this dilemma of *her* making, she says, is "....to look to cosmic consciousness as an *a priori* principle ....from which we may validly derive the appearance of phenomena by syllogistic reasoning." This is hilarious. How can you *derive* appearance from cosmic consciousness by syllogistic reasoning ? Appearances cannot be derived, they simply appear or don't appear. Our indefatiguable guide, Ruth Reyna, is left in a solipsistic un-Universe postulating unreality and then deducing from super-awareness the appearances that make up the subject-matter of science ! Surely this calls for a heavy employment of Occam's Razor, and further insistence on the Principle of Least Effort, not to speak of the role that simple sanity might play.

RAJENDRA PRASAD PANDEY AND THE TREK OF NON-EXISTENCE

Our third and last philosopher, whose work impinges on the

epistemological problems of science, demonstrates the strong tendency in Indian philosophy today to attempt (unsuccessfully of course) to avoid making judgments about concrete content or even relatively concrete content, preferring to remain within the province of deductive systems (or games). He is Rajendra Prasad Pandey, whose book *The Problem of Fact*,[62] is concerned with the problem of objectivity and subjectivity. It is his view that :

> . . the whole realm of Objectivity is . . understandable . . It is perhaps therefore that some philosophers prefer to attach the notion of understandability to the meaning of words—a view which, we think, is not essentially different from ours.[63]

Pandey then quotes Bertrand Russell's view that "the only thing you can really understand. . . . is a symbol, and to understand a symbol is to know what it stands for". Knowing Russell's amplitude and bultitude for perambulation, this may mean that the symbol stands for either a mental concept, a pseudo-material, object, or a perception that may or may not be considered to reflect reality.

Pandey's next observation concerns his belief that he has shown that "Subjectivity can be significant only as experience,"[65] which means, I take it, that subjectivity cannot be objective, hardly a sage remark. Contrasted to subjectivity is objectively, the experience of a faculty which implies otherness. Otherness is absent from subjectivity, he says. Thinking in these terms, we can now understand that it "is because of non-otherness that Subjectivity *qua*[66] consciousness must be considered only as significant and not as meaningful."[67] Since one of the synonyms of "significant" is "meaningful" it is perplexing to guess what Pandey here has in mind. Yet, warns Pandey, yet— ". . . . the character of the otherness of Objectivity can be annihilated because it is relative to consciousness."[68] Again, one is hard pressed to make any sense out of this. Without taking a breath at this recondite remark, he goes on to say that the possibility of annihilation does not mean that Objectivity itself ceases to be. The reason for this, he says, is that Objectivity cannot be reduced to mere nothingness (as much as all idealists would like). What a relief! because we have just learned from Pandey that Objectivity *qua* otherness *can* be annihilated. Perhaps if we remove the *qua*, which is hanging on like a barnacle, it will give objectivity a slightly stronger grip on reality. Pandey appears to face this ticklish situation with well-merited caution, for he says :

> Now here there is a very delicate point in metaphysics [in dialectics]: are we to regard the state of no Objectivity-as-otherness as the state of nothingness? No!..because a state of nothingness is a state of no-experience and no-understanding while that of Subjectivity and Objectivity is one of experience, and experience and understanding respectively.[69]

Wait a moment, Sri Pandey! If no one is experiencing or understanding extra-galactic nebulae, does this mean that they disappear into nothingness ? Worse luck for the astronomers trying to chart the stars in their courses!

But, says Pandey, "before taking up the case of nothingness[70] for further consideration, let us consider again *experience* with respect to Subjectivity...." Righto. We are game. When we again consider experience with respect to subjectivity we conclude that experience as subjectivity is a state of infinite freedom[71] where subject and freedom are the same. (To my way of thinking, Western subjectivists cannot hold a candle to the supermundane flights of the Indian metaphysicians, nor can the Chinese for that matter). Turning again to objectivity, which Pandey capitalizes as if it were a sacred word along with Subjectivity, we can discuss Objectivity-as-otherness and Objectivity-without-otherness. Taken as content, objectivity or pure Objectivity is simply the state of Existence. But Existence is Subjectivity. It appears also that Subjectivity is undifferentiated being, completely other than nothingness. "Therefore," concludes Pandey "nothingness is the unintelligible *par excellence*!"[72] On the other hand, he notes :

> As we have already seen..the finitude[73] of Objectivity *qua* nothingness is nothing but its own creation; it is due to the confusion of mind with Subjectivity; the former treating itself as identical with the latter, which is responsible for the situation of otherness.[74]

Reading this should remind us that it is far safer to refrain at the start from kiting checks, than to get more and more confused in trying to cover them up in turn : running from Chase Manhattan to Imperial and from Imperial to Grindlays. Just so we now find Pandey busy keeping track of so many nonexistences that he is likely to make a more serious blunder than his original one—which was to begin without the absolutely essential imprimatur of a material world. Think of thy presuppositions at the beginning, before the evil days come, and thou speakest consummate nonsense. But it is a marvel to follow him to the bitter and tragic end. He says :

> So, Objectivity itself creates and removes its boundaries..when the boundary is removed, then what we have in knowledge is the undifferentiated[75] infinite..[and hence] [t]o look for a meaning in a metaphysical statement..is simply absurd. However, metaphysics, in spite of its being unmeaningful, is an occasion of significant expression in language..[s]uch is the fate of Objectivity, and with it that of facts in metaphysics.[76]

At first it was "*significance*" that was not meaningful; now Pandey has added that metaphysics can be significant without being meaningful. Our wit can scarcely fathom this. Why is it my fate to collect virulently subjectivist books, both Indian and Western, that turn the natural and the real world into crapulous derangements? It must be the mysterious work of *karma*.

Indian philosophers of science, except the materialists, are all saying the same thing: that we cannot know the real world of material processes, and that hence we must rely upon subjectivity. Most of them end up in solipsism, the ultimate home of idealism and subjectivity.

VANKEEPURAM VERADACHVRI AND THE QUEST FOR SPIRITUAL REALISM

Non-academic treatises in philosophy of science may appear from time to time. They are just as interesting as those produced by university professors in that they reflect the views of the concerned layman, usually of some profession like law or business. These lay treatises are usually attacks on some aspect of science. I discovered Varadachari's *Concept of Matter*[77] in a new supermarket in New Delhi, where it lay among other spiritual books such as the gītā. Like various writers on the same topic, Varadachari believes that impure matter has disappeared, having been transmuted into the gold of mentality. Says

Varadachari:

> In this essay I have endeavoured to expound the fact that "matter" is not inert and unconscious, as it is commonly believed to be; that it is infinitely rapid pulsation of energy analogous to consciousness, shifting into new patterns under the compulsion of an immanent creative drive.[78]

This creative drive which Lucretius called *clinamen* is not much different from free will, according to George Thomson, a materialist mentioned by Varadachari. Other scientists besides Rutherford, quoted by Varadachari, are Magnus I. Puke, Joad, Jeans, Eddington, A.W. Osborn, and Paul Brunton. Besides relying upon the above authorities for corroboration of his viewpoint, he departs into the byways of psychic phenomena, using such exemplars as G.N.M. Tyrell, B.L. Atreya, J.B. Rhine, and W. Whately Carington. Some of the other thinkers he mentions include F.W.H. Myers, Henry Sidgwick, and Henry Butcher. It is the same Sidgwick, who is famous for his defence of common sense at Cambridge before Moore taught there, and a forerunner of Popper in the criticism of historicism.

In defending the existence of psychic phenomena, Varadachari says:

> As another instance of psychic communication may be cited—the case of Indra Devi. It is stated that, after her contact with Sri Aurobindo in 1949 Indra Devi began to have numerous visitations and visions, in the course of which she was inspired by Meera Bai, the renounced devotee of Marwar of the middle age [sic] and that as a result Indra Devi produced about a thousand poems, the first series of which was published under the caption "Suddhanjali". It is further stated that all these poems were reproduced verbatim as the songs heard by Indra Devi in her vision and that she did not compose them.[79]

Other eminent scientists quoted by Varadachari are Sir Oliver Lodge, who claimed to have communicated with his dead son who had been slain in the first World War. "Henry" [sic] Bergson is mentioned, whose views, Varadachari believes support Vedānta and Max Planck, which German is quoted as having said: "I regard consiousness as fundamental and matter as derived from consciousness." His reward was to have an Institute named after him.

Varadachari calls his position "spiritual realism." He regards it as a view which regards "matter, life, and mind as spirit."[80] In order to bolster this position, Varadachari quotes Hiralal Halder, who, as we have seen elsewhere, was a leading right-wing Hegelian in India before Independence. According to Halder as quoted by Varadachari:

> Mind and its object are the two opposed aspects of the one all-embracing unity, which is also mind. This all-inclusive spirit within which distinctions of every kind arise, which is bifurcated into subject and object, is the ultimate reality—the universe in its last interpretation.... We thus see from the highest point of view the Universe is spirit self-distinguished into the knower and the known, the cognising mind and its owner, namely the object-world that is cognised.

This thought of Halder, which originates in his "Realistic Idealism," (*Contemporary Indian Philosophy*, pp. 322-3), is to be expanded by adding the following coda:

> The material world is inwardised in mind and mind is externalised in matter. They are the correlated phases of the one all-inclusive spirit. In preaching this truth idealism *is in no way inconsistent with realism* [my italics—DR]. It does not make its business to deny the reality of the world. On the contrary, it strongly affirms it. It does as far as realism does but goes farther, maintaining that the world is indeed real, even what are called ideas as component part of it, but that in order to know that it is real it has got to have a mind.[81]

But what is the real? It is not anything that anyone can think. It is anything that is independent of thinking.[82] We can think imaginary numbers[83] and fantastic things.[84] That these have some basis in reality should not be forgotten, for example that mythology is based on ritual and symbolism, in themselves anchored in man's productive activity. If all consciousness were to vanish from the world there would still be left the objective, material, physical, independently existing world, including bodies or the bones of extinct idealists. The mind in the meantime, an activity grouping together a series of aspects of the composite activity we call thinking, ceases functioning with the body, as Lucretius correctly perceived two thousand years ago. We approvingly quoted the view of Igor Hrusovsky that:

> Ontology as a philosophical discipline, i.e., as the universal science of the *universal* character of Being, worked out on the basis of the most universal philosophical categories, can be no different, *in principle*, from those regional sciences which have as their object the characterization of qualitatively limited being. Philosophy and the special sciences together represent the total structure of knowledge. External, independent reality is the source of all concrete cognition.[85]

Matter and thinking are totalities of interpenetration "..and mingle reciprocally with one another; they are relational. We speak therefore, of the many-faceted structure of a particular concrete totality and of the multi-dimensionality of the whole of objective reality."

One of the problems that the Indian philosophers have (along with many Western European and American philosophers) with materialism is that their definitions are framed by idealists. For example, D.J. Kalupahana believes that nihilism is a form of materialism when at the same time he characterizes "..nihilistic theory as denying the reality even of physical objects."[86] It is clear that if nihilists deny the reality of physical objects then they are not materialists but idealists. Kalupahana cannot have a materialist both a believer in physical objects *and* a non-believer in physical objects. As Nāgārjuna (often cited as the greatest nihilist in Asian thought) says: "Insofar as I am concerned, those who speak of the reality of entities and who assign them distinct existences cannot be considered truly knowledgeable of the (Buddha's) teachings."[87] In other words, if you are not an idealist you cannot truly understand Buddhism! A materialist philosopher like Debiprasad Chattopadhyaya might respond by saying: "Such a person (materialist) understands Buddhism only too well!"

We must, lamentably, leave Varadachari because we are called to other writings of Indian philosophers concerned with the philosophy of science. We turn, then, to P.T. Raju, who has been deeply concerned with the evils wrought by science and technology.

P.T. RAJU AND THE PRE-EMINENCE OF SPIRITUAL NECESSITY

To fight against the evils of science and technology, one must have a plan of attack according to Raju. Towards this end, Indian philosophers should consider, he opines, that (1) Philosophy ought to be socially useful. We can only congratulate Raju on this discovery and implore him to hold that thought. (2) We philosophers must be on guard against the organization which follows from industrialization since it will lead to the weakening of traditional values (which Raju thinks that we should retain) and thus undermine the older philosophical positions and in addition enfeeble traditional religious positions. Again, we can only congratulate Raju in seeing that this is true although we cannot agree with him that this is also distressing. (3) Philosophers must be able to answer the charge that economic necessity is more basic than spiritual necessity. Raju is to be congratulated because of his experience in the West and his study of comparative thought which have made him aware of these problems. He stands almost alone among traditionalists in seeing the crux of the contradictions involved in India's change from feudalistic colonialism to modern capitalism. The Indians have a trump card by which to take the trick, however, according to Raju. Indian philosophers have" . . [t] he peculiarity of Indian religions . . that they have no dogmas, are not based upon faith, but on reason and experience."[88] Considering that every form of orthodox Hinduism bases itself on the alleged truth of the eternal vedas, no one can take this statement of non-dogmatism at face value. Other presuppositions of the Hindu religion such as karma, transmigration, and *moksa* are believed despite reason and experience. I do not mean to condemn these dogmas in relation to other religions which also have their own dogmas held contrary to reason and experience. Each religion with which I am familiar, is based upon dogmas that defy not only reason and experience, but also override the laws of formal logic (of somewhat less concern) whenever they appear to be inconvenient.

We must say, nevertheless, in Raju's favor, that he sees some of the problems arising from modern science and technology. It is his response to them that is irrational although understandable within the context of traditional idealistic Indian philosophy.

More complex in his approach and also broader in that he is more aware of the socio-economic vectors of modern life, is

B.G. Tiwari, who combines sympathy with change with a reluctance to condone it.

B.G. TIWARI: HORSE SENSE VERSUS SOLIPSISM

Tiwari, after agreeing with Lenin's rejection of subjectivism as an important contribution to modern thought with special reference to *Materialism and Empirio-Criticism*, attacks Lenin's view of matter as being "dogma," for matter which has the "..sole property of ..objective reality. "[89] This simply gives us, says Tiwari, "..the old mechanical notions of causality and substance, time and space, ether and electricity, etc. which reappear in a new grab."[90] Furthermore, Tiwari complains, the notion of matter in motion is retained as is the notion that mind is thinking matter. Thus mechanism is retained and naturalistic realism is intermingled with epistemological realism. Tiwari continues his lament that :

> Modern realists as opposed to Leninist realists, notably Moore, have effectively tackled the problem of perception and objective reality without attaching themselves to any dogma concerning "matter or Substance."[91]

What is this new notion of matter? Hertz says that matter consists of atoms; Becquerel says that the positive electron is a "new component of matter." *Van Nostrand Scientific Encyclopedia* says that :

> From the classical microscopic point of view, matter is that which can be considered as an aggregate of material particles. It is capable of being located in space and time and possesses the inherent property of *inertia.*[92]

The Penguin Dictionary of Science defines matter as:

> A specialized form of energy which has the attributes of mass and extension in *space* and time. They go on to define energy as The capacity for doing work. The various forms of energy, interconvertible by suitable means, include potential, kinetic, electrical, heat, chemical, nuclear, and radiant *energy.* Interconversion between these forms of energy can only occur in the presence of *matter.*[93]

It should be noted that most Western philosophers of science are fearful of even discussing matter let alone defining it. In the *Readings in the Philosophy of Science*, edited by Herbert

Feigl and May Brodbeck, "matter" does not even appear in the index. Instead, we find that Feigl, who wishes to be "free from metaphysical nonsence," introduces "*physicalism in the strict sense.*" Such strict physicalism, says Feigl, consists in "..postulating the potential derivability of all scientific laws from the laws of physics,"[94] Feigl claims that

> A more definite meaning may be suggested by formulating the thesis of physicalism in the following way: The set of physical laws which enables us to deduce the facts of chemistry will also be sufficient for biology and psychology.[95]

Thus has grown the incredible power of formalism! Now it enables Feigl to *deduce* "facts." One begins by "postulating the potential derivability" and ends with deducing "facts." There hardly seems to be any need for the inductive activity of science working with the real material world. Contemplation and pencil work [mathematics] are sufficient.

But let us examine just how Tiwari's hero, G.E. Moore attacks the problem of matter and external, independent objects. First of all, Moore believes in external objects and tries to set down a proof for their existence.[90] Even so, according to Tiwari, this does not commit Moore to a view of matter. We can deduce, however, from that fact that Moore believes that external objects are *not in the mind*, that they are *not mental*, and hence that they are physical or material. We disregard the transcendental third-way of Plato, Plotinus, and Popper (our transcendentalizing P's). What is broadly meant in one definition of material is that it is something that is not mental. Moore's proof for the existence of things outside ourselves (outside our minds) is as follows:

> I can prove now [says Moore], for instance, that two human hands exists. How? By holding up my two hands and saying, as I make a certain gesture with the right hand, "Here is one hand," and adding, as I make a certain gesture with the left, "and here is another." And if, by doing this, I have proved *ipso facto* the existence of external things, you will all see that I can also do it now in a number of ways: there is no need to multiply examples.[97]

What we have in our mind, like after-images, hallucinations, dreams and so forth, are not external objects which exist independently, according to Moore. What are such non-mental externals? Physical objects made of matter. Why all the mystification from Tiwari? Why not call a spade a spade ?, instead of employing epistemological gamesmanship ? We know the answer, of course. He is trying to make out a case for idealism. But Moore has already made this impossible for Tiwari in his essay "The Refutation of Idealism."[98] In that paper Moore states:

> The question requiring to be asked about material things is thus not: What reason have we for supposing that anything exists corresponding to our sensations? but, What reason have we for supposing that material things do not exist, since their existence has precisely the same evidence as that of our sensations?[99]

We can say, therefore that Moore first of all believes in matter by implication, and second of all that he believes in matter because he simply believes in it; Moore also puts forward, tangentially, another reason for believing in the existence of external material objects, but does not pursue the subject to its conclusion, namely that to argue in favor of or against the existence of external material objects requires that they exist as a necessary presupposition for the inquiry itself.[100] This is a powerful argument, but not pursued by Moore. Moore views matter as physical, external, and a thing. In what sense this is less dogmatic than Lenin's view I cannot imagine. (See Moore, *Phil. Studies*, p. 89, *Phil. Papers*, p. 139). It appears that among subjectivists, if you claim to know anything objective at all you are dogmatic. This subjectivist dog in the manger attitude is incumbent upon subjectivists (hence, relativists) since they begin with the presupposition that we can never really know anything but our own thoughts. (See Berkeley, Hume, Kant and Rickert).

In the next step of his discussion Tiwari mentions that Lenin does not and cannot prove that the mind is thinking matter. Indeed, says Tiwari:

> ..[c] onsciousness, in a serious epistemological inquiry remains an ultimate fact, behind which we cannot go. And the hypothesis that the mind, considered as a thing, or the physical fact, which determines the existence of mind, consists of the resonance energy of simultaneous discharges of living cells or some other parallel physiological theory remains, for the epistemologist, a pure and irrelevant piece of speculation, which has nothing to do with an inquiry that seeks to describe and evaluate the facts and characteristics of perception.[101]

First of all, what else can the mind be thought of except as a "thing," a function or a process? In German, *Ding* means thing, matter, creature, transaction—in any event something to which one can refer or indicate with his finger. If someone asks me where my mind is, will I point out to the planets or to my body? It can be thought of as a function, of *various things*. A function is a function of something else, which is a thing or collection of things. Second, the fi !dings of neurophysiology are always pertinent to epistemological discussions, and if they are *findings*, they are not mere speculation, since speculation is an attempt to get "findings." Finally, we should note that Tiwari becomes petulant even at the thought of scientific evidence being introduced into "philosophical" inquiry. It is the opposite view to that held by the neo-positivists that philosophical evidence should not be introduced into scientific discussions and who thereby imply that Philosophy is largely dead because its functions have been taken over by science, religion, or art. In this way Tiwari is carrying out the reactionary distinction made by K.C. Bhattacharyya. By reactionary, I mean the attempt to avoid the inclusion of all or any "scientific" knowledge in philosophical thinking. Such pointillism leads one to absurdities. Science may not be helpful in shoring-up dogmas of the revivalists of Hinduism or Islam, but it is necessary in working out a philosophical system. By dogma I mean thinking based on unalterable, eternal, formulae passed down from authorities, and accepted regardless of concrete conditions of space and time, historical and social conditions. It is employed by those who split truth into the higher and lower such that certain "truths" are seen to be immutable and above criticism and analysis. Dogma may also be defined as proof by quotation. Dogma's consequence is the stagnation of knowledge and practice. During the bicentennial celebrations in the United States (1776-1976) no official could bear the thought that the American Constitution and the Bill of Rights were not immutable, eternal and beyond discussion or question.

Worthy of consideration is that Tiwari's view of knowledge is egoistic. Knowledge for him seems to have no social dimension, for, as he says :

> ..*we must trust our own perception, before we can perceive the brain on the dissecting table.* If, then, the correctness of our per-

> ception has to be assumed, when we observe the brain, it may be, equally assumed in the case of other objects of our world. Knowledge is, thus, a unique phenomena [sic], a fact *suigeneris* [sic]. We may investigate nature, development, functions, etc. But it is not the business of the epistemologist to allow himself to be entangled in physiological inquiries and il-logical [sic] theories concerning its origin.[102]

We may, says Tiwari, investigate the nature, development, and functions of knowledge, but not the origins. "Perceptions," however, are socially conditioned. "Our" perceptions are the perceptions which have culminated as a result of a long evolutionary process. If what we perceive were always unique, then we could not have a public world. Nor could we have an objective science which depends upon a public world. The epistemologist, like any other philosopher, cannot know too much. He should get entangled in all the phases of knowledge that he has time to examine. How illogical theories have been generated by examining the origin of knowledge in physiological inquiry, Tiwari does not say. We cannot take seriously, then, his contention of the purity and "suigeneris"-ness of perception.

Let us return to Tiwari's charge that Lenin's acceptance of mechanism[103] ruins Lenin's position. What does "mechanical" mean here? It refers to that type of materialism which grew with rapidly progressing mechanics and mathematical physics in the seventeenth and eighteenth century in Europe, particularly Britain. Engels defines the mechanical view of nature as implying "Mechanics in the broader or narrower sense—known only quantities, it calculates with velocities and masses, and at most with volumes."[104] The point is that it does not deal with qualities, or if it does reduce them to quantitative considerations wherever possible. Another feature of mechanical or vulgar materialism is the belief expressed by its proponents that consciousness and social phenomena are the effect of exclusively physiological and chemical processes. Lenin, of course, does not agree with this position simply because it goes counter to a major presupposition of histodiamat: that man's consciousness is a social product. All psychical processes are governed not only by chemistry, physiology, physics and so forth, but also by man's social being. To believe this is to discard mechanical materialism, which Lenin did. On the other hand, if Tiwari wishes to find out more about mechanical materialism which he wrongly associates with Lenin, he should read the works of Vogt, Büchner, Moleschott, and Hobbes. Dialectical versus mechanical materialism, along with historical materialism, is a philosophical synthesis embracing social as well as natural phenomena.

"Lenin," says Tiwari, "no doubt, again and again emphasizes the fact that, for the dialectical materialist, the sole property of matter is its objective reality." The reason for Lenin's insisting on this is that matter appears in so many forms that to list them all would require a small encyclopedia. What all the forms have in common is the content of being objective reality, independently of being known. That is a brilliant generalization partially generated by Lenin's reaction to positivism.

Now that we have seen what "mechanical" means we can see that the following statement of Tiwari is simply false. He says of Lenin's dialectical materialism that it "..uses the old mechanical portions of causality and substance, time and space, ether and electricity, etc. which re-appear in a new garb."[105]

Tiwari's next move on his schedule of non-refuting refutations is to attack Lenin's representationalism and defend the view of Mach and Avenarius who say that we know only our sensations. Tiwari says that "Their theory, in spite of obvious imperfections, was an advance on representationism."[106] On the contrary, the most obvious imperfection of Machianism is that is results in solipsism, for as Ludwig Boltzmann pointed out, on this thesis of Machianism "..other people are only the sensations of the speaker."[107] Besides that, the external world disappears along with other people, which is another way of saying that you are alone with your sensations. Tiwari, undoubtedly, has a deep kinship with Chaudhury, the philosopher of science who defends transcendental solipsism. What makes Tiwari so delightful, in addition to this buffoonery, is that he poses to himself as a realist when he is simply another creeping solipsist. According to him:

> If, instead of denouncing the opponents for their characterising the materialists as "Metaphysicians," "Dogmatists," etc., who had made a "leap," a "salto-vitale" from the sense perception to the object, Lenin had thrown representationism over-board, as several modern realists [?] have done and if he had fully developed Dietzgen's view that the sensum is an aspect or a part of the object, he could, effectively, have established the validity, in its

> own right, not only of the thing-in-itself, but also the sensum.[108]

There is no gainsaying that. If Lenin had become a subjectivist and idealist holding that the sensum, or what was to be later called the sense datum, was part of the object, then his objectivity would have disappeared and he could stand out as a representative of anti-scientific cogitations and cohabit with the various positivists. He of course warned Gorki against doing just that, in the very book where his photograph appears showing him playing chess with Bogdanov! It was Plekhanov who had wrongly said, according to Lenin, that belief in the existence of the outer world is "an inevitable *salte vitale*.[109] Tiwari then rambles on to something which might be taken to be a conclusion.

> The strength as well as the weakness of Lenin's epistemology lies in its conviction that the method of ordinary commonsense and experience can be applied to the problem of philosophy. So long as it confines itself to giving an account of what is experience in the act of knowledge, it is on safe ground.[110]

As I try to read meaning into this, it seems to be praise for Lenin's horse sense, although Tiwari himself shows little horse sense by personifying epistemology such that epistemology itself *has* a conviction. "Look out for epistemology, as it is going to raise some ugly conviction or other !" he might as well have said. But we continue to quote more of the longer passage from which the previous quotation is taken :

> The trouble starts [not with Lenin's old horse sense but], when it [epistemology is loose again] attempts a metaphysical explanation in terms, which take us behind the ultimate fact of knowledge.[111]

Of course nothing can take us behind an ultimate because what we mean by an ultimate is that which nothing can take us behind. This is clearly a case of what Bernard Berenson called "metafussics." With barely a breath of hope left, we continue to quote from the long peroration to the affect:

> Then it [epistemology] begins to tread on ground, which is rather beyond the scope of epistemology proper. Even assuming that epistemology is entitled to attempt and [sic] a solution of these problems, we must note that there are those, who can discover no inconsistency between psychological realism and metaphysical idealism.[112]

Yes, Hilary Putnam puts this in proper focus, He says:

> When a realistically minded scientist—that is to say, a scientist *whose practice* is realistic, not one whose official "philosophy of science" is realistic—accepts a theory, he accepts it as true (or probably true, or approximately-true, or probably approximately-true). Since he also accepts logic he knows that certain moves *preserve truth*.[113]

Tiwari himself seems quite incapable of discovering not only such inconsistency, but very many inconsistencies. He was not really serious in his earlier criticism of Lenin. He was simply miffed that histodiamat philosophy (and psychology) turns out to be so clear and easy once you have cleared away the idealistic, positivistic, solipsistic nonsense which opposed it.

Tiwari is like a man on a mountainside walking through mist. For a moment, here and there, he sees far and clearly. But most of the time he is in a fog. He is correct in saying that "Lenin is right in repudiating subjectivism, the doctrine that the objects are the creations of the subject's fancy."[114] He is also close to being accurate in pointing out that M.N. Roy's "scientific materialism" is "..very close to the empirical and rational realism [read "idealism"] of Bertrand Russell."[115] However, as fascinating as all this is, we must move on to the work of Sengupta.

PRADIP KUMAR SENGUPTA AND THE SUBJECTIVIST STRUCTURE OF SCIENCE

Sengupta in his *Demonstration and Logical Truth* is primarily concerned with the problem of inference in general. Since his inquiry goes beyond what is normally called logic today, I include it under this account of "philosophy of science in India" admitting that it is difficult to make inflexible rules governing the division.

Sengupta's first point is that inference is different from perception because "[i] n inference the basis must be clearly apprehended as the cause or ground sufficient to justify our result. The ground must be such from which our result necessarily follows."[116] "Thereforeness," he says, must be present. That leads to the so-called paradox of inference which consists in this: that on the one hand we "induce" certain necessary conclusions from premises, while: "We agree that subjective novelty is certainly requisite, since there cannot be any apprehension of objective movelty without any change in the subjective standpoint."[117] This judgment follows Sengupta's discussion of the relative merits of the Keynesian view of subjective novelty versus the Millian view of objective novelty. Referring back to the last quotation, beginning with "We agree.." we see that Sengupta is likely to have the usual ontological problem of putting the cart before the horse, of having the subjective feelings determine the objective situation. Yet, as if anticipating our caveat, he says on the same page: "The conclusion of an inference must indeed be objectively new and still such a conclusion can be shown to follow necessarily from the premises." The so-called paradox is merely verbal, Sengupta says, for it depends upon the ambiguous use of "contain" in the phrase Y is contained in X.

Having examined Aristotle, Mill, Bradley, and W.E. Johnson as their work impinges on the problem of inference and induction, Sensgupta now turns to the inductive views of Carnap and Popper. Sengupta claims that "Carnap holds that one fact can never be inferred from another fact. From experience we can never infer anything lying beyond immediate experience. Since 'all' lies beyond 'some', from 'some' we cannot infer 'all". According to Carnap, inference is strictly deductive and induction is not a mode of inference at all he maintains. Therefore, says Carnap : "..induction on the basis of accidental properties is invalid and induction on the basis of essential properties is superfluous."[118] This may prove to be embarrassing. Does Popper help us out of this fix or is he going to give us more of his broth? Not really. Popper goes "one step beyond" by withdrawing even the methodological utility of induction. About this, Sengupta cheerfully says:

> If Carnap takes away induction from the province of pure logic and confine [sic] it to that of applied logic, Propper [sic] seems to shut all the doors of logic whether pure or applied. According to Popper, induction has nothing to do with logic. It is a matter with the layman in his wild goose-chase [sic].[119]

Carnap finds induction lacking in rigor and discipline. Popper declares induction to be a myth. This state of affairs is most unsatisfactory, says Sengupta, so let us return to the drawing-board—in this case, back to Bradley.

For Bradley "....the only purpose of logic is to set out the general essence of inference and to explain its merits and demerits."[120] There are, Sengupta finds, four major views in the Bradleyan formalism with regard to inference. They are :

(1) Logic is not an art of reasoning. It is a science in general and moreover an abstract science. It only endeavours to construct a general outline of the structure of inference. It does not attempt to furnish any true picture of the world.

(2) There is no limit to the number of terms and premises in inference. There is also no fixed order of the premises.

(3) The principle of identity which works at the background of inference is a purely formal one and has been sharply distinguished by Bradley from the metaphysical use of the same.

(4) The needed universal character of the premises is to be understood not in terms of the denotative and existential characters of judgements. It only shows the formal and abstract nature of the premises concerned. Every universal judgement is hypothetical.[121]

Having quoted this, Sengupta passes on to the next chapter without comment. He closes his discussion of Bradley by presumably agreeing with that worthy to the effect that "....the conclusion does not go beyond the premises and deals with elements numerically the same as and continuous with those of the premises. And so that the conclusion could not be a bare repetition of the premises,....gives a new connection [previously unnoticed ?—DR] between the same elements."[122]

Sengupta next raises the question of logical truth. For "truth", however, he substitutes "proof" which he claims he will explain in terms of necessity. Sengupta states that :

> The conclusion can be said to be proved when its relation to the premises is such that to accept the premises and deny the conclusion would be a contradiction. This is what we mean when we say that the conclusion follows necessarily from the premises. The conclusion can thus be said to constitute proof when it is logically necessary. This "logical necessity" is what we mean here by "logical truth."[123]

We have discussed "logical" truth earlier. A clearer way of

treating it is to simply call it *validity* if by truth we mean a proposition or sentence that corresponds with reality. Ontology and reality have been emptied out of logical truth. In short, there is no "truth" in logical truth, only validty. Putnam recognizes the same difficulty in Carnap's view of "confirmation".[124]

Sengupta now says that ". . . . the logicians want to conquer Thought."[125] He should say this about the positivist logicians. This attempt to "conquer thought" instead of reality or thought without reality, leads them to disregard natural phenomena, yet, Sengupta says :

> . . they do not object to any extension of the use of inference as is done by scientists for which the scientists are eternally grateful. Furthermore, the logicians intend to arrange the scientific discoveries in strict logical order so that they can determine the logically certain procedure (necessity). But unfortunately Nature is too capricious to follow the rules of logical stringency and Thought is too liberal to abide by the dictates of the natural phenomena.[125]

Nature is terribly complex and anarchic. Thought just wanders around in a quandry. But let us remember, says Sengupta, ". . the honourable caution of Hume[126] that past is not guarantee for future."[127] If that is so, then we have two limitations : the complexity of nature and the possibility that the future will not resemble the past which prevents us from "strict empirical certainty," although we may be *logically certain*. Sengupta, it appears, accepts the notion that ontological and logical considerations are separate and separable, although it is not quite clear how far he is willing to move in this schizophrenic direction which seems to rise from drinking the Popper's broth and imbibing Carnapian soma. Sengupta does believe that it is best for the logician "to avoid the question of truth."[128] That leaves him with "formal" validity which rules out statements about the real, material, world which are true, false or decidable.

Sengupta does make a contribution, besides his dialogue itself. It comes at the end of his book. He says that entailment should not be thought of in terms of material implication, but in terms of logical implication. Solomon Levy had pointed this out to me in 1952 when we were sitting near the Gate of India in Bombay having an evening discussion. Sengupta, like Levy, is correct although Sengupta may be largely unconscious of the major reason for its soundness. There is no *material* or *objective* reality involved in "material implication," and therefore it should have been called, from the start, *ideal implication.* It perhaps would have been if Bertrand Russell had got as clear about ontology as he was unclear about marriage. That a false proposition implies any proposition can only be interpreted idealistically. It is an implication seen as stripped of all real material considerations, hence all sensible, empirical, encumbrances. So-called "material implication," insofar as induction is concerned, leaves us up in the idealistic air because there is here no connection between logic and the material world. It was such considerations that led Russell into "....the field of pure a priori thinking" to finally descend "....into the actual world of experience" as John Passmore points out in *A Hundred Years of British Philosophy.*

Two years after the appearance of the work we have been discussing, Sengupta published *The Logical Structure of Science* (1970). In the first chapter, Sengupta proposes that science be confined to natural phenomena so that when he uses the term "scientist" he means natural scientist. As he avers,

> It is essential for the scientists that the vast world of objects be studied without any prejudice or preconceived bias[129] because any such bias may affect the scientist's attitudes and thereby contaminate his entire study.[130]

Studies in the sociology of science or the sociology of knowledge (often described as Marxism without Marx) have shown how it can be overcome, indeed how it is overcome. In his account of the scientific enterprise, Sengupta quotes with approval C.J. Ducasse, Popper, Duhem, Bertrand Russell, J.O. Wisdom, Carnap, Cornelius Benjamin, Hempel, Kneale, Nagel, J.S. Mill, and Bunge. Of these, perhaps only Ducasse's view may be regarded as realistic, so it would appear that the shaping of Sengupta's opinions or "temper of mind"[31] will be overwhelmingly idealistic and subjectivist. Ernest Nagel, in his discussion of physical reality, appears to have both a realist and an idealist proclivity. At first the realist seems to dominate. Then follows the idealist who talks about "the reality of scientific objects." Gutzmore, of course, pointed out that there are no such things as "scientific propositions" and more than that there are no "real scientific objects." It is clear from the index of his *The Structure of Science* (1961), that when Nagel writes about "the physical" he never mentions "the material," but the elision leaves him in idealism. For example, this is obvious when he says: "....ironically-minded instrumentalists can acknowledge the physical reality or existence of many theoretical entities."[132] If theoretical entities exist in the sense of physical reality, in what sense do non-theoretical entities exist ? We seem to have stumbled on Platonism and Popperism again.

Sengupta claims that it is the task of the scientist to provide a systematic account of the facts and phenomena of nature (excluding, as we have seen, non-natural science such as economics, history, philosophy, psychology). On the other hand, there are legitimate metascience and methodological studies about the sciences and about scientific procedure he believes. Methodology consists in studying scientific method in which inference is not strongly stressed. Metascience should be the study of the role and structure of scientific statements and inference, Sengupta holds. To provide "....a unified logical framework of the theories of science....using logical rules and principles....as well as....metalinguistic principles such as Pragmatics, Semantics and logical syntax over and above the pure logical rules of inference,"[133] is the main task of metascience. "The metascientist will introduce unity or disunity and logical structure into the results of the various natural sciences." To eliminate the claims of sophistry, "....[t] he utility of such a study is to make....a systematic interpretation of the scientific methods enabling us to understand better the results of the sciences."[134] Another way of putting this is that methodologists draw out the methods that the scientists use often without analysis. Although there is always the danger (?) that scientists will pay little attention to the methodologists which seems to be the case with the works, for example, of Kuhn. When several different methods are presented, the methodologist has to decide which one be favors in terms of its logical cogency"....in keeping with the spirit of actual scientific explanations." It may be that subjectivist (idealist) philosophy of science is to the actual process of science what theology is to the actual process of religion—namely ritual. There is only the demand of the coterie of barnacles and parasites. The reader will note here that there is an emphasis by Sengupta not on rigor, but on "the spirit" of the scientific enterprise. Since the scientist may be unconscious of what he is doing, presumably, the methodologist brings this to light in conformity with "logical cogency."

We can here reflect that we have the possible bias of the scientist compounded with the bias of the methodologist who must make value judgements about the various methods used. Since the methodologist in India or Western Europe, in any case, is almost always a subjectivist or positivist, this bias appears to be extremely glaring to anyone holding materialist or materialist-realist presuppositions (such as Moore, R.W. Sellars, Bernal and Hilary Putnam, [when he is wearing his materialist hat]. Perhaps the effort of a third investigator is required such as a sociologist of science, who tries to bring to light what is unconsciously employed by the meta-scientists and methodologists. Such would be the Bulgarian philosopher Dabrin Spassov who has pointed out that materialism does not follow from science but that science presupposes it ". . . . any scientist presupposes the existence of objects of knowledge independent of the process of knowing them. . . ." ("On Quine's Epistemology and Philosophy of Logic") *Proc. of the XV World Congress of Philosophy*, (Sofia, 1973) III, 390. Throughout this chapter and the preceding one, I have assigned myself to this role. A fourth investigator, needless to say, is encouraged to examine the biases of the third investigator.[135]

Besides metascientific and methodological considerations, there are other problems, according to Sengupta. For, as he says,

> It is very difficult to determine the range of scientific data; it is equally difficult to say what methodological features a discipline must exhibit in order to be a science. We have seen that Popper, Kuhn, and Lakatos failed at this task.* What a philosopher of science can at most do is to presuppose the present framework of the sciences as interpreted by whom?—idealist or materialist? and draw out their common features both as regards content and method. Thus the philosophical data will include every fact of science and every interpretation of such fact if it is taken in the relationship of "explained" to its "explanation."[136]

But the very agreement about what is a "fact" will depend upon the metaphysical-dialectical viewpoint of the philosopher of science! This is not to say that no facts are objective, only that which ones are objective have to be determined by social practice which is then recognized as relatively binding. Objectivity may be provisional in the sense that science is an open enterprise. It is not provisional in the sense that "anything goes" or "anybody's view is as good as anyone else's". What we say the facts are is determined by our presuppositions as well as by the philosophy of science that we have.[137]

Sengupta, presumably nearly totally innocent of such considerations, paraphrases Carnap at this point to the effect that "....one's criterion of testability is a matter of free choice."[138] But this is surely arbitrary if not downright silly. It goes counter to the social practice upon which science is based. It is not a free choice, but dependent upon past scientific practice and agreement as well as some modicum of present input. It is a choice determined by past and present scientific practice and agreement. Hence testability cannot consist in mere logical coherence, but must get at the objective references so that the methodologist is able to operate outside the mental squirrel cage. Furthermore, objective reference depends upon *correspondence* and not *merely coherence*. When C.D. Broad writes of "epistemological" and "ontological" objects in *The Mind and Its Place in Nature* he is setting the scene for making this distinction. Such a distinction, argued in good faith, seem to miss the crucial consideration, as also does the viewpoint of modern Thomistic realism, that the social dimension is crucial. It is not simply one man against the epistemological and ontological world, but a host of men *working together*, *co-operating* that makes the ontological separable and knowable. Descartes gives us the erroneous impression that it is he, God, and the object, which make up the total *dramatis personnae* of knowledge. The positivists have eliminated God; the Thomists have kept him. But they both ignore the remainder of mankind and its social history. And Descartes' dreams are the "offspring of indigestion" rather than a typical experience of mankind, as Leonard Louis Levinson suggests. And in our day, there is a third difficulty, to be found especially in French thinkers, the cultivation of "....an anarchic 'disorientation of the senses', following the example of Rimbaud...."[139] This is important because it shows that the anarchism and subjectivism of positivists are simply part of a larger intrusion on materially based contemporary good sense. The final reference depends, not merely on coherence, but on correspondence, as the logical positivists, logical empiricists, subjectivists, and other idealists have yet to discover.

When we turn to Sengupta's chapter entitled "The Paradoxes of Confirmation" we discover that :

> ..generalisations supposedly produced by non-deductive inference are verified in the sense that we can never verify more than a finite number of instances. But such generalisations are falsifiable in the sense that we may find out at best one contrary instance in view of which any said generalisation has to be accepted as false.[140]

In addition, he notes that inductive generalisations are only approximations, but never absolutely true. This failure appears to give deductionists and formalists like Sengupta a kind of thrill akin to wiggling a loose tooth. It is the one thing that they can say to put down scientists who are changing the face of the earth and heavens while deductionists and formalists are grinding their teeth impotently in their rusty squirrel cages. Sengupta, along with his mentors, never gets down to the questions as to whether there is an external world or whether we know it, how it corresponds to the statements of the scientists, and so on. In short, he writes like Carnap and the other logical positivists, flitting like bats in the vacuum of formal relations, or, as Montaigne put it, involved in "....sharpe-witted subtilties....without substance, about which Philosophie doth often busie its selfe."[141] An exemplar of this is Sengupta's paraphrase of Carnap's view of the nature of the logical relation which the concept of inductive probability is alleged to involve. Says Sengupta :

> In discussing the nature of this logical relation Carnap warns us that the probability value (whatever it may be) does not concern the evidence from which we start nor the hypothesis which we arrive at. It concerns only *the logical relation* between the evidence and the hypothesis.[142]

Instead of getting out of the squirrel cage we keep revolving in it foreover, now observing a perch in relation to a sliding door, and now noticing that the bars of the cage are parallel to each other! It seems that we shall never pay attention to what is outside the cage so long as we can play with icons, symbols, hypostaties, ideal nexuses, operations, hypothesized elements, and so forth.

We are disappointed in Sengupta's work. It does not bring to light the logical structure of science at all, but the logical structures invented by Carnap, Popper, and other logical positivists who must still discover *the source* of their icons and symbols before they can say much sensible about science.

C.T.K. CHARI AND HINDU RIVIVALISM UNDER THE GUISE OF PHILOSOPHY OF SCIENCE

No other Indian philosopher writing on the philosophy of science demonstrates more subtly the current contradictions which the Indian intellectual faces when trying to develop a *modus vivendi* between his appetite for contemporary science with a tenacious grasping of past dogma. Chari says of himself that he :

> ..has examined questions centering round relativity and time, quantum physics, neuro-physiological hypotheses about a "brain-mind liason" and the alleged "revival" of "buried memories" by the deceptive hynotic technique.[143]

With Chari we note a cloud of quotation marks around any terms that he would like to leave unanalyzed so that he is not committed to any obvious metaphysical or ontological position. In the article from which the above quotation is lifted, Chari warns us that the naturalist critic of Indian spiritualism cannot with justice demand verification *on his own terms*. One must be willing, he states, to disengage oneself from the *trivialities of sense-perception* when one proposes to test spiritual truths.[144] I find myself impatient with mindless empiricism until mindless and matterless spiritualism begins to challenge it. What this expresses is Chari's unmediated dualism. Therefore, he is not going to give priority to scientific method, although he continues to write like a dedicated philosopher of science. This is a bit disheartening when one reads the impressive list of titles of his articles on scientific topics. Despite this initial[145] disappointment, however, I feel impelled to give an account of the writings of Chari to see whether he turns out to be a solipsist along with Chaudhury and others whom we have discussed. Abdur Rahman is concerned about the social isolation of scientists, another problem tangential to solipsism. See his *Science, Technology and Economic Development* (Delhi : National, 1974), p. 41. From the above statements we may deduce that Chari is neither materialist nor naturalist. If not, he is an idealist, since being a non-idealist realist would entail his belief in either naturalism or materialism, since we believe that the sobriquet : "Platonic realist" is a contradiction. It is like being a Dostotevskin optimist or a Rousseauian rationalist or a Marxian capitalist.

Since the time that Chari wrote his doctoral dissertation, entitled : "On Some Spatial Representations of Time," we can say that he is the only Indian philosopher with a wide reading in physical science who writes extensively in the philosophy of science. It is said that his work on time should be considered a contribution to the philosophy of science. In "A Note on Multi-Dimensional Time"[146] he claims that the various forms of multi-dimensional theory of time can be extracted from Dunne's hypothesis, one part of which can be preserved. Although I recall having heard Dunne mentioned in a seminar in metaphysics in early 1940, I remembered nothing about him To refresh my memory I took down his book from the university library shelf where it was last taken out six years ago, and I turned to the summary. It proved to be a defense of what Dunne called "Serialism." According to Dunne,

> Serialism discloses the existence of a reasonable kind of "soul"—and individual soul which has a definite beginning in absolute Time—a soul whose immortality, being in other dimensions of Time, does not clash with the obvious ending of the individual in the physiologist's Time dimension, and a soul whose existence does not nullify the physiologist's discovery that brain activity provides the *formal foundation* [my italics—DR] of all mundane experience and of all associative thinking.[147]

We do not need to write a gloss on this statement. Dunne's next statement is less worthy of quotation. It is that it is the nature of this soul and of its mental development to provide us with a satisfactory answer to the "why" of evolution, of birth, of pain, of sleep, and of death.[148] Dunne finishes his book by claiming that this soul in time gives us the possibility of telepathic communication. The part that can be preserved, according to Chari is Dunne's "....infinity of dimensions" as "the demand for *very general* topological representations of multi-dimensional time."[149] But an *hypothesis* is not a *demand*, it is a tentative explanation. Chari goes on. Between two chosen instants of clock-time, one may indicate linear time. Resulting subsets may be denoted by *sigma* : the class of subsets corresponds to the epochs of multi-dimensional time. It is possible also to indicate subliminal time as well as clock-time with reference to ultimate time. Chari warns us, however, that :

> The extreme uncertainties about the interpretation of the symbol *t* in certain newer quantum-mechanical theories should warn us that we may have to contemplate departure from familiar notions of time-ordering even if we cannot give concrete interpretations of all the symbols.[150]

Ah well, what matter! The main thing is to get time sufficiently mixed-up with soul and other conceptions such that no one is quite sure what he is talking about.

Chari next examines in his "On the 'Space' and 'Time' of Hallucinations,"[151] the view of J.R. Smythies which states that hallucinations are spatial and colored entities. After a brief survey of relevant literature, Chari asks the following questions : "On the basis of what measurements and conventions I ask, can the 'metrisability' and 'dimensionality of the 'space' of mescal visions be determined ?"[152] Things are "desperately vague" says Chari, "when we turn to intricate mescal hallucination which have lattice, cobweb, tunnel spiral shapes, there were also alterations in shapes such as polyopia, dysmegalopsia, dysmorphopsia. All objects may appear to be situated at the same indefinite distance; or, they may recede into space without altering their size."[153] There seems to be "distortion" of time which becomes, Chari says, "unstable" when a subject is under the influence of mescaline. Chari raises the interesting question : "How, on the basis of the empirical data adduced by Smythies, can we rule out the possibility that some hallucinatory spaces 'are *not* 'length-preserving' and 'angle-preserving' "?[154] He further remarks that "Considering that geometry itself in any accepted sense may fail for certain problems in the nuclear domain, we may well doubt whether a multi-dimensional space will satisfy the needs of a still young and unsure science of psychology."[155] Since what we have got is a multi-dimensional continuum ever since the general acceptance of space-time in relativity it is difficult to know what Chari is driving at here. Either one has multi-dimension space or uni-dimensional space. Does anyone claim to have the latter today ? What is the point of his remark ?

Another of Chari's excursions into time is entitled "Philosophic issues about Irreversibility in Classical and Quantum Physics." [155]He concludes after nine pages of historical summary-for it must be admitted that he usually goes through a great deal of material—that:

> ..it should be noticed that no statistical interpretation of either classical or quantum physics has quite eliminated the basic reversibility or symmetry of laws in space-time. [Chari is here referring especially to thermodynamics. But enough doubts remain for us to be too completely satisfied with either of the two solutions distinguished recently by O. Penrose..and L.C. Percival: Latropic irreversibility is simply inconsequence of the size and complexity of the system under consideration and our ignorance of.. initial conditions; (b) Non entropic irreversibility is the effect of "The rest of the universe" on the "open system" we have chosen.[156]

Chari might also have mentioned reversal processes in chemistry and biology. But we continue with this rather long quotation:

> Both the entropic and the non-entropic temporal asymmetry seems to yield at best a "local" irreversibility holding for limited, though very long, periods. Is a philosophic generalization possible by an inversion of our standpoint? May not the uncertainty about "initial conditions" and about "influences coming from infinity" themselves be rooted in a basic temporal asymmetry? It may not be subjective time that is flowing in the positive direction of L. Brillounin's..observation-parameter but observation and inference that may be bound up with a time characterized by a march of events (or "becoming") in a privileged direction which is "statistical unexpectedness ("future").[157]

It is difficult to know what is meant by "influences coming from infinity."[158] Although in projective geometry it is postulated that lines which are parallel in Euclidean geometry meet "at infinity," to speak of influences "coming from infinity" is to speak as if *something postulated exists out there* waiting to compel or stimulate us. It is perhaps plausible for writing science fiction and for apologists to bring in non-scientific "evidence" and "principles" of this sort.

In "Information, Cosmology and Time," Chari is concerned with the relationship of information theory to all large-scale studies of the universe. After his usual sortie with the relevant literature available to him, he concludes that:

> The anisoptropy [the exhibiting of different values when measured along different axes] of time should be examined not only in

> relation to the alleged "recession of the galaxies" or the possible "creation of matter," but in the language of "making observations and securing information about very large regions of the universe."[159]

Again, Chari verges on science fiction in merely mentioning creation of matter, because if matter is "created," someone or something does the creating. Admit that, and we are back in revivalist land. Some of the research along the line of anisotropy has been conducted by J. Neyman, R.L. Scott, G.C. McVittie, and D. Layzer. Chari notes that ". . . .[i] n Jordan's new cosmology, the gravitational constant varies with cosmological time and matter emerges in a concentrated form as supernovae."[160] Of perhaps greater interest to philosophers of science is Chari's suggestions that "Any *adequate* statistical decision about the "uniformity" or "non-uniformity" of large-scale processes may mean a full-blown cosmology."[161] I find this an exhilerating suggestion, the realization of which would make it mandatory for any person writing in the area of philosophy of science to work out an ontology of some sort, than which nothing is more evaded. Information theory, says Chari, may assist us in clarifying what are adequate and what are inadequate stochastic (stochastic refers to a process in which the outcome is believed to have some initial probability) models in describing natural processes.[162] I could continue to analyze Chari's fecund writing, but as fascinating as this might be, I should only continue to demonstrate that India has produced one philosopher of science who *prima facie* is doing creditable work in certain areas of this specialization. That he does not tie up philosophy of science with humanistic considerations indicates that he is caught up in the routine of individualistic specialization so common in the West. Nevertheless, I think it instructive to learn what Chari more or less openly espouses in the realm of metaphysics and value theory as these impinge on the scientific enterprise.

Pursuing that line of analysis, we find that Chari in "Human Personality in East-West Perspectives," mentions that W.E. Hocking "opened up for me a vast spiritual universe, a whole enchanted sea lit by the softest dawn and trees swaying to the rhythm of whispered words. . Nobody can study Hocking without an enhanced respect for the mystical experiences of the human race, without the realization that nothing is so superficial as mere philosophical sophistication."[163] Chari also realized as a result of his encounter with Hocking that "It is de rigeur [sic] for a mystic to be a tentative realist."[164] Yes, the "vast spiritual universe" seems to alternate between the material and spiritual for Chari. It is nearly impossible to determine whether Chari is or is not a realist, although I know that he is not a naturalist or a materialist. He seems at times to be sympathetic to neo-Platonic realism, but then quickly changes direction and quotes in succession a number of authorities who are giving idealistic interpretations of the material universe. What other clues have we? One is that Chari never quotes a clearly identifiable naturalist or materialist, preferring always to cite positivists and other idealists. He quotes Hocking, for example, as saying in 1930:

> Various systems of psychology may be worked out, internally consistent, mutually inconsistent and all inadequate. If we characterize the real as a self, we express the unknown in terms of a commensurate unknown, as an *infinitude* of the same order.[165]

Perhaps this is the origin of Chari's special infinity from which things impose themselves on us. If we assume that anyone could ever know just what it is Hocking is writing about in the above quotation, for example that "the real as a self" signifies anything, Chari comments upon it with great courage and then drops the subject. He says: "I submit that the prophecy is coming true."[166] What prophecy? Hocking has written a conditional, not a declarative sentence. If it is a prophecy, does it mean that in the future (after 1930), various systems of psychology will be worked out but inadequately? It takes no seer to predict this. Again, is the real a self? And if it is, is this what Chari means by being a "tentative realist?" Certainly the view is not realism, but Personalism. Is Chari a Hocking Personalist? We are left with these questions unanswered, so move on to another topic—to Chari's examination of psychical research.

Regarding this fascinating field, Chari says that "It seems to me that Hocking refuses to take the bold step which C.D. Broad and C.J. Ducasse have taken."[167] Evidently Broad and Ducasse think that reincarnation is a "valid possibility" thus mixing up formal validity with material truth. Chari, girded for his foray into "possibilities in the infinite," says:

> I have elsewhere argued that Indian Yoga is both mysticism and realism. In an early article, Hocking regarded mystical experiences as true when taken as "psychology reports," like income tax returns can be taken as "psychological reports" rather than descriptions of reality?, but false and misleading when taken as "metaphysical statements." I suspect (I may be wrong) that Hocking, like other Occidental and [Oriental—DR] thinkers, shrinks from having to activate Maeterlinck's "paralysed eastern lobe of the brain." Perhaps this inhibits his approach to psychical experiences....A biographer of St. Francis Xavier, after expressly warning readers that miracles are two a penny for the credulous, refers to some well-authenticated incidents in the life of St. Xavier suggesting a possible precognition as well as a possible clairvoyance.[168]

Just why would Chari accept as "well-authenticated," incidents mentioned in a life of St. Francis Xavier written by J. Broderick S.J.? Obviously Chari does not take lightly the claims of psychic phenomena. Yet, he cautions us on the next page, by saying: "A mysticism in which doubt and conviction merge needs a watchful empiricism."[169] Preferrably a mindless empiricism? The idealist and tentative realist in Chari merge as do doubt and conviction.

Chari goes on to quote my old friend, William F. Goodwin of Wisconsin, to the effect that "Indian mysticism is morally dead, since it does not enjoin 'work in the world' but withdrawal from the world." To which Chari replies: "I do not suppose that Hocking, with his richer and wider experience of the East, would agree."[170] As a matter of fact, outside of a visit in 1931, Hocking scarcely set foot in South Asia, East Asia, or Southeast Asia. This information comes to me from Hocking's son, Richard.

My own view is that Goodwin is correct. In addition, I will add to what he says, trained as he was in the now non-ontological agora of Berkeley Philosophy that Indian mysticism was ontologically dead even before it was morally dead. Chari's last statement of concern to us is that "Mystical personalism in the East and in the West, I would insist, has an infinite concern for human beings."[171] On the contrary, it seems to have a strictly finite concern for human beings other than the subject. It seems to have great concern for the individual trying to achieve *mokṣa* rather than troubling itself with the plight of the mass of mankind. This infinite concern[172] seems, on the basis of actual consequences, leads to "infinite" non-happenings in the areas that might illustrate these concerns: in agriculture, education, science, medical care, industrialization. Most concerns of the mystic have led to results best described as approaching zero. Such concerns in India have led to increased misery among the common people.

Chari's mysticism, his revivalism, his religiosity, his cheerful ignorance of and indifference to, social science, inhibit sound work in the philosophy of science. He has an active mind interested in many facets of mathematical theory but it desperately needs to be channeled by historically-viewed scientific method and disciplined by conditions laid down by recent social evolution. Furthermore he may need to be brought out of social isolation into the ongoing practical work of reconstructing Indian society. Part of this is to inspire confidence in real, living, socially beneficial science.

KRISHAN LAL KWATRA AND THE CRISIS IN PHYSICS

Krishan Lal Kwatra has made a notable contribution to Indian philosophy of science in his paper "Philosophy of Microphysics."[173] Discussing the crisis in physics, Krishan Lal exhibits at once that he is not a scientist who believes "..any nonsensical theory that may be put forward in the name of science," as Christopher Caudwell, paraphrasing Max Planck, stated in his *Crisis in Physics*[174] in 1936. It was never far from Caudwell's mind that the crisis in social relations in the boom and bust of the 1920's and 1930's throughout the capitalist world was reflected in the crisis in physics. As he puts it, "science..gives information about non-social reality. But it gives such information *to* society. The knowledge is conscious knowledge. It has therefore to be cast into the categories of society."

Krishan Lal attempts to overcome the alleged indeterminacy attributed to Heisenberg's work in physics. Heisenberg's Principle of Uncertainty, however, was not that the activities of nature are uncertain, but that the position and velocity of an electron or elementary particle can never both be exactly known. Only approximate figures can be obtained if both are calculated together, although either one independently or separately can be known to any required degree of precision. As the precision of one variable increases the other decreases. The connection between them is the tiny quantum of action called Planck's constant. The wave function, then, is not a description of the behavior of particles in reality, but a description of limited information. "In other words, quantum mechanics, although capable of describing what we know, is not a description of reality"[175] says Krishan Lal. It is significant that he writes as if reality can be known, instead of suggesting that what we know is to be treated *as if* it were reality, as in the Kantian ontological mode of description. Then what is a wave function a description of it not a description of reality? It would certainly seem that it is some sort of a description of part of reality. Krishan Lal goes overboard in implying that if it isn't a complete description that it is no description at all.

Krishan Lal quotes Einstein's letter to Popper (1935) to substantiate his belief. Einstein said in that letter that the so-called indeterminacy in physics should not be interpreted in the usual fashion since if through the position and momentum co-ordinates it is possible to predict something, "then this something must exist in reality."[176] In other words, what was being predicted was not merely in someone's mind, but was independent and objective. In pursuing this problem of indeterminacy or so-called indeterminacy, we are left with the following alternatives according to Krishan Lal : (1) Either we have only a half-way scheme which needs to be supplemented in quantum mechanical description, or (2) we have an admission of the impossibility (at present) of precise description of all systems looked at individually. The first view is that of Heisenberg and Bohr. The second view is that of Einstein, Vigier, and Bohm. K.L. notes that "Einstein's life long [sic] opposition to the usual interpretation of quantum mechanics was based on a profound belief in what he regarded as the intelligibility of the universe." Every modern materialist shares in this belief. Like Einstein, he cannot believe that "uncertainty forms the basis of nature."

It was De Broglie in 1927, says K.L., who first recommended, not a probabilist indeterminacy outlook on quantum mechanics, but a "concrete and deterministic interpretation."[177] De Broglie, as we have seen elsewhere, gave up this view and later joined Heisenberg and Bohr under the pressure of seeing in the chaotic world an irrational element particularly noted by German and Austrian scientists after they had lost the first world War. Around 1950, however, Blokhinshev and Terletsky pointed out that it was unnecessary to accept the Heisenberg-Bohr interpretation. Bohm in 1952 attacked indeterminacy in principle because it sets a limit to the precision of all possible measurement, thus legislating for future science. Popper and his friends legislate for future science in their discussions of verifiability, recommending that nothing be accepted unless it is verifiable in advance. Heisenberg and Bohr assumed that the usual interpretation of the quantum theory was not merely a result of the current stage of development, but was an absolute—something which they had claimed to be avoiding. Krishan Lal makes no mention of an interesting side-effect of this viewpoint which was to make many Christian priests and ministers (those who bothered to learn anything at all about science) gleeful because, since everything was indeterminate, the only hope was to rest in the Christian faith so as to have a stable mooring. Thus were they able to generalize from this specific problem in physics. Von Neumann (1955), according to K.L., went even further than Heisenberg in the direction of indeterminacy by asserting that "..there is hardly any warrant to assume a deeper level which would ever permit its exact determination." K.L. rightly calls this nothing but capricious generalization. Of von Neumann's view, A. Lande said in 'The Case for Indeterminism" (1955) that von Neumann:

> ..first introduced the whole involved apparatus of the quantum theory, thereby lifting us to a highly exalted plane. In this lofty atmosphere he carried out certain calculations, then brought his results back to earth again—when he could have stayed there all the time.[178]

It is almost a relief to find that the physicists are doing what the subjectivist philosophers are doing. It makes the social world all of one piece. By abrogating reality, they jack themselves up into a transcendental sphere which substitutes (badly) for reality.

Krishan Lal is well-attuned to the flavor of subjective idealism that pervades the field of quantum mechanics. As Bohm put it: "..the conclusion that there is no deeper level of causality than von Neumann's is just a piece of circular reasoning, since it will follow only if we assume beforehand that no such level exists."[179] It is important, K.L. warns us, to not overlook these disagreements because they are fundamental to one's view of reality. One cannot escape the (dialectical) relationship and operation in which decisions are made about physics (and reality) by pretending to avoid philosophy. Most scientists who avoid philosophy are simply subjectivists following the crowd of idealists who have a more authoritative style. The style quickly turns into the content of their thought, the content having been pumped out by an ingenious idealistic pump. Krishan Lal might have added to his discussion that what the physicist decides about reality is important in other areas as well—it has social implications. For if indeterminism "holds" then there can be no human planning and hence no socialist rationality—which is what the Poppers of the social sciences determined to fight against as early as 1919. Physics is shot through with ideological warfare just as are sociology, psychology, anthropology, literature, and philosophy. There is no neat, isolated world for the student of mechanics any more than for the student of philosophy of science. The whole of discourse, from the humanities to the "hard" sciences is a battle ground littered with idealistic corpses. Ignorant and rigid physicists (unrubberized?) have contributed to the cause of reaction, anti-science, and the stoppage of needed social change. Many physicists have been in the forefront of the reactionary movement intended to bring about the dematerialization of matter.

According to Krishan Lal there is contingency as well as determinacy, for he says:

> ..what is contingent in one field becomes necessary to another. Thus the initial position and velocity of a given object may be determined by varous factors which operate naturally in the given environment such as wind and waves or by agents who may set these conditions of on their own accord. However, in turn, these large fields have their necessities and contingencies. The wind and wave follow certain dynamical laws but are contingent in the sense that the initial conditions indicated through these laws depend on factors such as movements of objects in space, ions from the sun, etc. which are outside the field of their applicability.[180]

But K.L. may have too narrow a view of what the field of applicability is. Ultimately it is the whole universe which makes up the field.

About the random movements of the electrons : they are only apparently random. Every rational scientist assumes this. Freud, despite the idealistic system of explanation that he developed, had the belief that psychical states were not random, but that they followed regularities within the field. In this sense he was a scrupulous scientist. Leonardo da Vinci in his flying experiments believed that the flight of birds did not depend upon random qualities, but that there were regularities that could be applied to human use if only man could discover what these regularities were. The Wright brothers and others believed him and went on to eliminate the random factors.

Krishan Lal looks with particular disfavor on such statements as the following made by Heisenberg:

"The conception of the objective reality of the elementary particles has thus evaporated in a curious way, not into the fog of some new, obscure, or not yet understood reality conception [the reader will note that Heisenberg has unobtrusively switched from an objective to subjective mood by changing "reality" to "reality conception".[181] This same trick was used by John Dewey to change "reality" to "experience"—DR], but into the transparent clarity of a mathematics that represents no longer the behavior of elementary particles but rather our knowledge of this behavior."[182]

Pertinent to this subjectivism of Heisenberg is K.L.'s comment that "..if randomness is equated with irregularity then how can it be governed by *laws* of probability without involving a contradiction?"[183] Precisely, Krishan Lal. How indeed? The viewpoint of indeterminacy with its treatment of "absolute contingency" as the meaning of randomness is closely linked with the belief in "completeness' of theory. And such a belief necessarily leads to ruling out as meaningless and irrelevant any attempt at finding new fields wherein individual events can be accounted for, Krishan Lal shrewdly points out. In such we have the metaphysical, or better, "conceptual" basis of subjectivism in physical science, which spreads like cancer throughout the whole intellectual body—from the social sciences to the humanities and finally attacks even necessitarian theology which itself has been "damaged" by 20th century physics—quite inadvertently one would think. K.L. does not make this point himself, but his work may be used as a basis for understanding the insidious reactionary thrust of subjectivism in physics and the philosophy of physics.

So, despite our earlier discouragement in reading much of Indian physical-science oriented philosophy of science, we are considerably cheered by Krishan Lal's appearance on the otherwise rather bleak horizon. Here and there in the darkness there flickers a light.

Yet we must not forget that much the same problematic holds in the Americas, Africa, Western Europe and selected parts of Eastern Europe, especially Poland which was, of all Eastern European countries, most influenced by positivism and mindless empiricism. It is not much different in the British Commonwealth.

PHILOSOPHY OF THE SOCIAL SCIENCES

Of the social sciences in India, political economy may well have advanced more than history, psychology, or sociology. Archaeology and anthropology have sometimes moved ahead with alacrity also, bound as they were to progressive British activities. Briffault had his influence on Mortimer Wheeler, whose world-wide fame rests on his heroic work at Mohenjodaro and Harappa as much as on anything. Wheeler said that "we are all in some sense Marxists at heart, just as we are all Newtonians or Darwinians. Historical materialism, by whatever name we call it, has interpenetrated the fabric of our studies.." Such was the case with his admirer D.D. Kosambi, one of the Indian luminaries of the twentieth century. But before we approach the views of the Indian Engels, we turn to the valiant effort made by Susobanh Sarkar to help clear the stables of the social sciences.

SUSOBANH SARKAR AND THE FIGHT AGAINST MYSTICISM IN THE SOCIAL SCIENCES

Up to this point we have been mainly discussing philosophy of science as it impinges on natural or physical science. We now turn to the rare instances where philosophy of science is applied to the social sciences in India.

Sarkar is trying to lay down the requirements for scientific history, choosing as a departure, an examination of Arnold J. Toynbee's philosophy of history as exhibited in his *Study of History*. Sarkar, first of all, praises the courage and diligence that produced one of the sweeping histories of all time. He epitomizes Toynbee's achievement in a little over three pages, which I shall reduce to half a page.

> First, the society of Western Europe plus North and South America has been expanding throughout the world for 400 years. As this expansion is now ending, it raises the most critical issue of our time.
>
> Second, Russia (U.S.S.R.), which developed centralized state power to protect itself from the West, is ready now for counter-attack, using the materials borrowed from Western civilization.
>
> Third, the Muslim world, instead of borrowing freely from Western civilization, confines its imitation to military weapons, and hence is confined in its capabilities.
>
> Fourth, Hindu civilization was attacked and conquered, whereas the Russian (Soviet) and Muslim societies were never thoroughly subdued. Despite this, Hindu society remained aloof from Western civilization, since here the difference was more fundamental than between Western Europe, Eastern Europe, and the Muslim world, since the latter were the offspring of the Graeco-Jewish source. Hindu civilization is entirely different.
>
> Fifth, Far Eastern society is also free from Graeco-Jewish influence. It was attacked in the 16th century. This failed, but it was again attacked in the 19th century, when it succeeded.
>
> Sixth, when an attacked society sets up its self-defence, some rays penetrate deeply and others shallowly. Technology penetrates deeply while religion only shallowly. Technology from the West is far less important than its religion which has not penetrated deeply. Thus we have the traditionalists in Russia, the Wahabit or Mahdist agitations in Islam, and the Gandhi Movement in India.
>
> Seventh, the prognosis for the future (for Toynbee) is that reactions today are very like those between the clash of Graeco-Roman society and the world outside. The solution is a new religion of brotherhood like the worship of Mithra, Isis, Buddha, and Christ—a new religion that states that salvation lies in forging personal links to a living god. This is popular mass movement. The ruling class, at first aloof, "joins" the procession. In that event, we will find an end to the present upheavals.[184]

That, according to Sarkar is the main thrust of a *Study of History*.

There are also secondary considerations. One of these, says Sarkar, is Toynbee's alleging that the West, since it has rid itself of oppression, is not concerned about oppression in "Russia". Sarkar points out that this is simply false for the West has not gotten rid of oppression. World (capitalist) unemployment is a major example of this. Says Sarkar: "The real reason for alarm about the U.S.S.R. is the fear of an unfamiliar social system and not any distaste for oppression."[185] Sarkar makes this appear to be sheerly psychological instead of emphasizing the fear based upon a realistic appraisal of the U.S.S.R. as a competitor in a world searching for raw material and markets. He is right in maintaining that "The defence of democracy is not an inspiration here, it is merely a make-believe."[186]

Sarkar points out that Toynbee believes that the Western world cannot accept communism which is a deviation from Western ideals and therefore inadmissable. Here Sarkar replies by pointing out that "..in actual history, liberalism, democracy, socialism and communism are integrally linked together in the West's life. They are successive stages in the fight against the older orders.[187] Toynbee also points out that communism is an adversary, not only of Western ideals, but also of the ideals of Hinduism. His argument is that communism is a distorted "ray" of Western civilization and therefore cannot enter deeply into the totally different Hindu society. But, as Sarkar points out, the other incompatible rays are entering with no difficulty: technology, secularism, parliamentary "democracy", capitalist industrialism. On the other hand, if the communist ray cannot easily enter, that is a proof of its far greater value, according to Sarkar. In addition to that, if the reception of communism in Eastern Europe is possible because of the Graeco-Jewish atmosphere, it must also be welcomed in the similarly conditioned Muslim world. He adds to this : "And what about the acceptance of communism in China which is outside the Graeco-Jewish orbit?"[188] How does this square with Toynbee's religious interpretation of history?

Sarkar thinks it odd that Toynbee finds so much to admire in Mustafa Kemal's Turkey, in a Turkey where Western civilization has been assimilated, and where a veritable "renaissance" has transpired. Toynbee considers the Turkish "upheaval" to be the greatest revolution in history! It appears to be also one of the greatest secrets in world history, particularly since for anyone visiting Turkey, it would appear that Westernization is far from complete. Noteworthy, however, is Turkey's entrance into the NATO armed camp. It may have been this which persuaded Toynbee that Turkey was moving along at breathtaking speed. In addition to this Turkish thesis, which by 1985 seemed inordinately weird, Toynbee produces another one hard to swallow. This one is that the Jesuit fathers were erecting Christianity suited to the East, when unfortunately, internal feuds among the Christians developed, so that "..the Christians themselves blasted the noble prospects".[189] These are two minor difficulties symptomatic of Toynbee's major ones. Among historical materialists Toynbee seems to be riding a tricycle among sport cars.

In Toynbee's favor, however, it may be said that he made many readers "curious about historical discussions." He dealt a blow to some kinds of piecemeal history which had limited investigation to minute details with attendant unawareness that some view of the whole is needed for understanding of the part. This is precisely what Thomson did for early Greek philosophy and Marx for nineteenth century political economy.

Also on the debit side, is Toynbee's fault of using each society as a living body—an organism. (After Darwin, the philosophy of organism developed, reaching its peak with Morgan, Alexander and Whitehead, and among orientalizers, Archie Bahm). It is likely, it seems to me, although Sarkar does not mention it, that the reason for Toynbee's choice is that he looked for a pre-histodiamat model of analysis, and this he finds in biological evolution. But can society be conceived in terms of birth, development, decay, and death—a "society" which begets children? Toynbee really ducks even the Darwinian model when he retreats to religion as the essence of culture. He returns to the Church Fathers. Toynbee has shown little awarness of the "science of culture", dependent on economic and technological grounds, and hence rejects the distinction in Europe between the feudal middle ages and the capitalist mode of existence. The great changes in Russia in 1917 seem miniscule to him because some semblance of orthodox Christianity still lives in the interstices of its society. Economic institutions such as slavery, feudalism, capitalism, and socialism appear to be relatively meaningless to Toynbee whose outlook may be compared thus to Augustine.

The general outlook of Sarkar has been described by Bishnu Dey, who says:

> The genuine rebel or revolutionary temper grounds itself in a classic mentality, in firmness which does not care for artyness or current fashions of taste, a firmness which does not drive out simple passion, just out of fear of being charged with sentimentality, nor does it forsake the ordinary or the familiar in search of the mirage of technical stunts.[190]

ARTHUR KOESTLER AND PSYCHIC MYSTERY IN SCIENCE

Sarkar, born to a Brahmā-Sāmaj family grew up disobeying the dictates of orthodox Hinduism. This makes him particularly interested in the work of Arthur Koestler, whose quasi-philosophical popular work has had impact on conservative and liberal intellectuals alike. I was again reminded of this when a former friend of mine, a student of philosophy and medicine, submitted how strong was Koestler's political influence on him.

According to Sarkar, it was Koestler who initiated the intellectual Cold War by publishing *Soviet Myth and Reality.* This would be difficult to prove. I prefer to think that the phenomenon began in various Western European and American countries at about the same time. Sarkar likens Koestler's "The Sleep-walkers" who are supposedly scientists, to H. Butterfield's *The Origin of Modern Science*, rather than to the larger canvas of J.D. Bernal in his epoch-making *Science in History*. Sarkar chides Koestler, who makes a show of stydying the history of science, for not mentioning Bernal in his bibliography.

Although there is "The abiding value of Koestler's survey (as a) lively sketch of the changing concepts of western cosmology (from the Greeks to Newton).."[191] there are grave defects in his work. Few will appreciate a number of pronouncements made by Koestler on the history of science, according to Sarkar. One of these, especially for Eastern readers, is the statement that the Buddhist outlook in contrast to that of the Pythagoreans is "negative" and "socially sterile". One of the great achievements of Pythagoras, Koestler maintains, is that Pythagoras tried to fuse science with religion: he says: "the Orphic ecstacy of his day was transformed in the Pythagorean world-view into a deep spiritual emotion."[192] It is hard to see how this is a contribution to anything but confusion. Along the same idealistic line, Koestler also claims that the abandonment of helio-centrism was a "failure of nerve," as a result of a lack of intellectual courage, thus using the same vague psychologistic non-explanations too often used by George Sarton and M. Rostovtzeff who also too often "explain" socio-political events by referring to "the spirit of the Greeks" and suchlike.

After Platonism had held sway for a millenium and a half, with its emphasis upon other-worldliness and absence of "dirty" science, experimentalism with the hands, says Sarkar, a thaw appeared. According to him Platonism was replaced by the "devious" Arab Gulf Stream which brought back Aristotle. Aristotle's scientific conclusions, unfortunately, soon froze into dogmas which created a demand for a return to the freer speculation of Platonism. This is Koestler's narrow view of Arab science. Neither Koestler nor Sarkar tells us why. Platonism, Sarkar paraphrases Koestler as saying, had encouraged the split of the universe into celestial and terrestrial spheres, geo-centrism, circular motion, and the conviction that bodies are naturally at rest so that their motion depends upon external compulsion or propulsion. Lynn Rose, an American historian of cosmology, however, says that Platonism did not encourage the split since Plato believed that activities in the celestial sphere could affect those in the terrestrial. Whether Plato believed, as it is alleged that Aristotle did, that there is only circular motion, is difficult to say, according to Rose.

After both Platonism and Aristotelianism were replaced by the new cosmology, the world was conceived as a machine, says Koestler. But the cosmology of the 16th century still did not obviously surpass that of Heraclitus, Aristarchus and Ptolemy, until Galileo shattered it with his telescope. Sarkar says that "Koestler's outline of the evolution of cosmology is admirable, but Koestler[193] has failed in his attempt to deepen our understanding of the historical process." I cannot concur is this judgment, but believe that Koestler's account is superficial. Sarkar believes that Koestler's account is not historically helpful because of certain shortcomings which include the following:

(1) Koestler fails to settle the central question as to whether historical explanations of mental evolution are a fruitless quest.

Significant connections must be made between man's mental and physical evolution (through technology, for one thing). Koestler speaks of the "zig-zag course of science", and the "devious, nay crooked ways" of the progress of science and of "the immense dark lowlands extending between the peaks" of achievement.[194] According to Sarkar, Koestler's comments "imply a certain lack of appreciation of the very nature of evolution. . a want of dialectical insight,"[195] which is to say a lack of knowledge of the interconnections, or to put it in another way, a plethora of positivistic squint-eye. Sarkar maintains:

> That the course of evolution does not run in a straight line, that advances and retardations are quite normal features, that contradictions are inherent in the process of development, that the rate of change is always uneven—these are after all fairly well-known facts, as the celebrated formulation of Lenin on the nature of evolution pointed out long ago.[196]

Examples of these would be useful to make Sarkar's account more understandable.

Koestler also gives considerable credence to the belief that important break-throughs in science are attributable to great men, and "other features" and "less determined by economic challenges and social pressures," although Keostler has admitted elsewhere that theories are shaped by social pressures.[197] Koestler's view of the contributions of great men is more like Emerson's than Plekanov's, in short, a psychological-spiritual Koestler also admits that account "[o] ut of a number of ideological mutations a given society will select that philosophy which it unconsciously feels to be best suited for the need." In his Epilogue to *Soviet Myth and Reality* he holds the view that concepts thrown up like mutations are established by natural selection adapted to the intellectual milieu.

Why, asks Sarkar, cannot Koestler make up his mind about social determining factors? Sarkar explains this anomaly as follows:

> The true source of Koestler's hesitancy can be traced to a resolute almost superstitious refusal to seek the roots of his "mental climates" which are made to loom like mere chance entities. His real objection is of course to the "economic challenges" which may underlie the social pressure.[198]

Yes, but can't we make one further step in this analysis. Why stop with Sarkar's explanation which is also merely psychological? Why not go one step beyond to the realization that what prevents Koestler's getting down to economic roots is that his life-work depends upon his ideological pose which is certainly conditioned by his knowledge. His very existence depends upon his ideological role as chief of the Western secular writers against socialism. As such he brings to bear every trick—religious or semantical—that he has learned in that cradle of mystification and obfuscation—Vienna. Vienna has harbored for 150 years the anti-humanistic sword pointed at many forms of scientific and humanistic Sacher-tort concern (except in music). Even the Sacher-tort is overrated, and like Viennese ideology, dried out. Koestler is to secular Cold War literature what Wittgenstein is to philosophical literature. They are both bold masters of evasion, subjectivism, and contempt for science and materialism. The advantages of these attitudes in the bourgeois world Koestler had long discovered and Wittgenstein was just beginning to discover. Wittgenstein, however, seemed to have been uneasy about his positions.

Acknowledgement that there are necessary connections between the economy and superstructure would be dangerously close to the histodiamat outlook and therefore piously avoided by Koestler. In one passage, Koestler slips into admitting that there was a play of material factors which quickened the pulse of change during the early Renaissance in Italy, but in this Koestler is no more fundamentalistic than the Catholic existentialist thinker, Gabriel Marcel, who wrote in a moment of lucidity: "..it would certainly be rash to deny *all* validity to the Marxian analysis.[199] "Much more remarkable", Sarkar notes, "is the avoidance of any comprehensive attempt" by Koestler to make "transparent connections".[200] Koestler provides us with, in avoiding the obvious, the light which is not dim except in the eye of the non-seeing beholder!

(2) Koestler's queer analysis of the "scientific mind" and the "process of discovery" is Sarkar's second caveat. According to Koestler, the scientist is like a "sleep-walker". Says Sarkar:

> Like Columbus, Kepler, we are told, reached America in quest after India. Furthermore, [t]he unavoidable effort in scientific analysis which is involved in a discovery is clouded over in [Koesterian] veils of mysticism.[201]

If discoveries are sometimes accidental, or appear to be by chance, they are certainly not supernatural, as Koestler implies. Furthermore, the tale Koestler himself unfolds in the book contradicts his mystification. (In his will he has left half a million dollars for psychical research). Presumably tangible causes and forces are always in operation, many if not all of which, are dependent upon being noticed by the scientists who are uncommonly awake and undogmatic. We know that in the United States, for example, the two largest religious groups (Roman Catholics and Baptists), who account for one-half of the total church membership, produce almost no eminent men of science. The reason for this is that adherents are carefully nurtured away from critical thinking and openended views.

(3) Koestler's third failure, and the most despicable one, is his surrepitious "attempt to pass off as the most acceptable social ideal a synthesis between faith and reason, religion and science. This raises the basic question of the feasibility of such synthesis in the light of historical experience,"[202] says Sarkar. Beginning with the so-called Pythagorean synthesis, this involved, Sarkar states, "..[n]o more than a vague emotion coupled with science as an immature adventure."[203] Following the so-called Pythagorean synthesis appeared that of Thomas Aquinas which contains the same recognition of science permitted by the Vatican when it rewrites the Darwinian theory of evolution to fit its unscientific (theological) presuppositions. Sarkar notes in passing that when Thomas Aquinas was faced with the same issue confronted by Koestler, he was more forthright and honest, for Koestler never tells us whether the scholastics were right or wrong in their final judgment regarding faith versus reason. He must imagine himself to be an Edgar Alan Poe writing a suspense-filled tale but without an ending.

If, as Koestler claims, science is leading the world to disaster and destruction, then what is the cause of this condition, asks Sarkar? Isn't it that vested interests encourage this misdirection of science for their own ends? And if this is true, how much of this misdirection is directly or indirectly tied up with the religion which Koestler is so anxious to keep? Isn't religion a useful tool used by the exploiters to keep the majority uncritical and in a mood of acceptance as if what the exploiters plotted was the will of God?

Sarkar points out that Koestler on the one hand (1) claims that science is not objective truth, while at the same time (2) maintaining that religion should synthesize with science. But why should he bother to combine these so-called different points of view, these different approaches, if both of them yield merely subjective opinion? Is it to make life more variegated and colorful for us to be presented by various untrue views which are equally but colorfully subjective? Surely this presents science as merely decorative.

Sarkar concludes his analysis of Koestler's outlook on science by a shrewd comment that moves perhaps a step beyond Stebbing's *Philosophy and the Physicists*,[204] the bourgeois complement to *Materialism and Empirico-Criticism*, where she demonstrates the absurdities into which idealistic physicists have recently fallen.[205] According to Sarkar:

> It is amusing to find that "disproving" the truth of science is immediately by a slight of hand turned into a demonstration of religion as truth. As soon as an adequate explanation of the nature of matter eludes us, an Eddington claims straightaway that the "mindstuff" is the true reality, or as Jeans describes the universe "a great thought" rather than a great machine—forgetting the glaring non-sequitur involved. It is far preferable to arrive at such "truths" by the traditional path of hoary philosophical idealism than to trot them out as the conclusions of modern scientific activity. . [a]s for the "crisis" [the crisis in physics] itself, clear thinking is imperative. The inability to define matter in a cogent manner cannot logically mean that matter itself "disappears", or ceases to exist. . [a]fter all, scientific truths are being verified in practice every minute, and a higher criterion of truth than practice is not available to us.[206]

It would have been better if Sarkar had not made this unnecessary concession to idealism by even hinting that there is an "inability to define matter". Clear thinking, as he has already said, [plus knowledge] can give us a decent enough definition of matter of such a provisional nature as to allow for future discoveries. If a new particle is discovered it is not necessary to go into a tizzy and mystically suggest that suddenly everything has changed and we no longer know what matter is. We actually know one more thing about matter, namely that it may be recognized in the form of a new particle (neutrino, quark). Hence de Broglization (dematerialization) is unnecessary.

Having examined Sarkar's analysis of the perplexities raised by Arthur Koestler, we turn now to Sarkar's view of Frederick Engels as an historian set forth in Sarkar's "Historical Writings of Engels.[207] In this illuminating essay, Sarkar first challenges the contention that Engels departed from and distorted teachings of Marx. The foremost argument to allegedly establish this is that Engels was primarily mechanical in his outlook. Sarkar demolishes this argument by quotations from *The Holy Family*. Engels maintains that man does nearly everything and that history does nothing, hence how is it that he can be considered mechanical in his treatment of social life? Neither Engels nor Marx believed that the superstructure was a mere reflex of the productive base or infrastructure. Says Engels "[W]hile the material mode.. is the *primum agens*, that does not preclude the ideological spheres from reacting upon them in their turn, though with a secondary effect.."[208] Along the same line, concerning the development of Germany, Engels says: "Without making oneself ridiculous it would be a difficult thing to explain in terms of economics the existence of every small state in Germany, past and present, or the origin of the High German consonant shifts.."[209] I do not intend to reproduce all the quotations, says Sarkar, to prove that Engels was not a mechanical materialist, yet I want to quote another passage to correct an idealistic misunderstanding that exists in "..the fatuous notion..that because we [Marx and Engels] deny an independent development to the various ideological spheres which play a part in history we also deny them an effect upon history."[210] For Engels, the economic condition is not the sole cause; neither do economic conditions produce an automatic or mechanical effect. When men make their own environment, they do not make it out of whole cloth, but use what is given to them. Nevertheless, the economic relations are the golden thread which runs through the historical-environmental conditions. If we removed all ties between ideology and the economic sphere, we would be left with nearly pure abstraction. For, Engels notes:

> ..if you plot the average axis of the curve, you will find that the axis of the curve will run more and more nearly parallel to the axis of the curve of economic development the longer the period considered and the wider the field dealt with..[211]

There may be some scholars unfamiliar with the terminology of mathematics who will mistakenly believe that this statement indicates Engels' mechanical propensities. But they should consider that both Engels and Marx as scientists, quantified whatever was susceptible of quantification. This did not mean, however, that they were ever unaware of the importance of quality which they repeatedly defend against the simplistic views of mechanical materialism. Nevertheless, Marx, for example, showed continous interest in mathematics from 1858 to 1883, the year he died. In a letter to Engels in 1863 he writes: "In my spare time I do differential and integral calculus....Aside from knowledge of the common algebraic and trigonometric stuff no preparatory study is needed except acquaintance with the conic sections."[212]

Engels was also fascinated by pre-history which is well-demonstrated in his study: "The part Played by Labour in the Transition from Ape to Man," written in 1876, but first published in 1896. According to Sarkar :

> Engels found the keynote of the process [of man's evolution] in labour—'labour created man himself.' All anthropoid apes can stand erect, but only in a most clumsy fashion. Man alone gained the habitually erect posture by acquiring and transmitting over the generations the skill of using the hands in manifold kinds of labour; thus, "the hand had become free" and with it had come the firm stance on the two legs alone. The skill in the use of hands in labour could alone produce the first tools.[213]

With tool-making comes correlative growth in other parts of man's organic being. While it is true that some large birds use rocks to open eggs or crabs to eat, and chimpanzees and other primates use sticks for tools, these are found by them as artifacts of nature. They did not manufacture them.

Not only was Engels unfairly attacked for mechanism, but also because he leaned so heavily on the work of Lewis H. Morgan, the pioneer American ethnologist from Rochester, New York. But it should be remembered that the discoveries which outmoded some of Morgan came after Engels' death. Even today we find opponents and defenders of Morgan arguing the latter's status, both in bourgeois anthropology and in anthropology based on, or congruent with, the materialist interpretation of history.

Sarkar makes the point that Marx and Engels were attracted

by the implications rather than the specific conclusions of Morgan's pattern. Engels had written that "..Morgan rediscovered in America, in his own way, the materialist conception of history that had been discovered by Marx forty years ago.."[214] Morgan reconstructed the prehistoric foundations of written history. So this provided a materialistic interpretation of prehistory as well as history, which V. Gordon Childs has pursued in his various studies. Sir Mortimer Wheeler mentions in his autobiography that every serious archaeologist is now somewhat of a Marxist, in the same sense that learned men after Newton became Newtonian. And he says also,

> Let me make it clear, then, that I would apply to our study of the Roman provinces to a far greater extent than hitherto, the whole machinery of modern materialistic archaeology.[215]

Engel's work in *The Origin of the Family* is praised by Sarkar for giving new and vigorous account of the rise of the Athenian state and the fall of the West Roman Empire. Concerning Athens, Engels traces the rise of the hereditary nobility and monarchy and the slavery that had started to come into the social order to ultimately become institutionalized. Only one thing was missing to seal general public recognition of the death of the old gentile system of the tribal order. Father-right and the inheritance of property by patriarchal families undermined the gens. The state had been born. Of this account Sarkar rightly says: "No more succinct concise account of the rise of a state, with some definite historical evidence, is indeed available in the range of Marxian classics."[216]

Engels was not only attracted to ancient history, but also to the history of Germany. Here he provided three seminal works, the last being unfinished: *The Peasant War in Germany* (1850), *Germany*: *Revolution and Counterrevolution* (1851-52), and the *Role of Force in History* (1886). His second book is still the short but penetrating account of the subject of the 1848-50 "storm in Germany". In *Force in History*, Engels develops his method in :

> masterly fashion. The letter to N.F. Danielson 17 October 1893 (Selected Works, II) is almost prophetic. It tells the Russians that, in the absence of the social revolution in the West, the institution of the mir is doomed; that differentiation in wealth will

destroy it as surely as rich peasants destroyed Athenian tribal society.[217]

Sarkar's final assessment of Engels is that "the extent of historical knowledge and depth of insight which we find..fully bear out the claim that he was one of the best educated men of his times...."[218]

D.D. KOSAMBI'S SCIENTIFIC APPROACH TO MACRO-HISTORY

Another Indian historian who worked on a much larger scale than Sarkar, and was in his own way the Engels of India, was Damodar Dharmanand Kosambi (1907-1966), whom I have elsewhere called the father of scientific Indian history. His masterpiece in this area is *Ancient India : A History of Its Culture and Civilization.*[219] His many-sided knowledge here is orchestrated into an exciting and authoritative study on the development and significance of Indian history up to medieval times. D.H.H. Ingalls, Sanskritist and Editor of the Harvard Oriental Series, says of him that even though "Kosambi was a Marxist; I am an anti-systematist and by nature conservative," that nevertheless Kosambi was a man with an "..instinctive respect for facts..I would almost call it reverence. He had the patience to weigh on a jeweler's scale each new lot of punch-marked coins that came into his hands.."[220] At the beginning of *Ancient India* Kosambi characteristically writes: "It is doubtless more important to change history than to write it.."[221] This reminds me of the statement that for the landlord it is more important to know his tenant's philosophy than his state of affluence. One is tempted to observe of Kosambi's comment on history, that if no one wrote history there would be little theory to encourage us to change it. Some insight into his encyclopaedic interests may be seen from an incomplete list of his publications provided me by Mr. Jerry Drost.[222] This excludes his writing in Sanskrit, Pāli, Hindi, Mārathī, and the Slavic languages.

According to Kosambi, India is uniquely interesting because of the continuity of its culture, surpassing that of Egypt, Mesopotamia, Arabia, the Romans, Incas, Aztecs, or the Yunnan people of China. Despite this, India has virtually no historical records. This serious deficiency can be partially compensated for in a reconstruction through archaeology, numismatics, ancient literature, and the surviving customs. "History," says Kosambi, "is the presentation in chronological order of successive changes in the means and relations of production."[223] At the base of all explanation is food production and the surplus production that determine the relations of production, relations which are embodied in the activities of people. Kosambi cautions the reader that :

> Our position [is] far from mechanical determination, particularly in dealing with India, where form is given the utmost importance while content is ignored. Economic determinism will not do. It is not inevitable, nor even true, that a given amount of wealth will lead to a given type of development..the complete historical process through which the social form has been reached is also of prime importance.[224]

For those particularly interested in linguistic philosophy, it is significant that Kosambi points out that "language itself was formed out of the process of exchange, new goods, fresh ideas, and corresponding new words all going together." Major changes occur in language when an old line of production is drastically changed from a primitive to a more advanced method. Consider the change from horse and wagon to automobile. Think of the words about horses and wagons that went out of use. My father could identify horses, parts of horses, wagons and coaches and their parts, the names of which, I, growing up among automobiles, never knew. The new form may lead to linguistic alterations as a result of increased population, which in itself develops changes in the superstructure not easily adapted. New linguistic forms emerge from this conflict. The language of histodiamat causes anxiety and confusion among bourgeois philosophers who are still living in the horse and buggy stage of philosophy. The bourgeois philosopher hears the word "dialectical" used in many new senses and contexts. If he knows nothing about calculus or higher mathematics it becomes more difficult for him to understand the language of the new philosophy and science. Superannuation occurs throughout history. The introduction of the crossbow developed a new language, introduction of the stirrup into Europe changed the vocabulary of millions of people, as Lynn White, Jr. has ably pointed out in his *Medieval Technology and Social Change* (1962). It is possible that old forms are broken by revolutions in the guise of reformations. If the class defending the old form wins, then "stagnation, degeneracy, or atrophy" set in as we well know in the 20th century from the example of Spain, the last holdout against parliamentary democracy (bourgeois democracy) in Western Europe. We see the same decline in Haiti, Chile, and the United States (at the onset of the second Great Depression (1970-72). The U.S.A. establishment tends to see all Third World revolution with socialistic elements as anti-democratic (communist), as Girilal Jain notes in the *Times of India* [*World Press Review*, February 1984, p. 44]. According to Kosambi, this factor helps explain India's early maturity and helplessness against later foreign invasions where adjustments did not occur or were instituted so slowly that they were too late to modify the main events of their time. The stagnation of China was only broken up by foreign invasions, replied to by the Boxers, the Sun Yat Sen supporters, and finally the Socialists and Maoists. Chiang was an agent of stagnation with respect to government.

Kosambi says that since there are so few historical documents in India, the reconstruction of its history must depend upon other evidence. The best way to collect this evidence is through the reconstruction of rural and tribal society. This attempt was begun by him in the countryside around Pune, where he often lived. T.H. Morgan attempted the same kind of thing among the Iroquois nations of Western New York State, a century before Kosambi. Among other things, such a reconstruction enables us to explain the "idiocy of village life."[225] Most of the remaining study by Kosambi is aimed at this reconstruction. By way of introduction to this task, he examines the forms of primitive life and prehistory, describes the primitive survivals in the means of production as well as the superstructure. But he avoids the approach of Pareto who was concerned with the "residues" that appear as "manifestations of sentiments and instincts just as the rising of the mercury in a thermometer is a manifestation of the rise in temperature."[226] The first class of residues the Pareto examines he calls "Class VI," or the sex residues based on the "instinct for combinations". This instinct "Prompts to combinations in general, for reasons which are feeling momentary, undetectable."[227] Pareto has something in common with subjectivist Indian philosphers of science when he can glibly discuss the "undetectable". He is like the university dean who exhorted his faculty to "prepare for the unforeseeable." Kosambi, needless to say, will not and cannot pursue this line of analysis since there are so few explanations based on instincts. The word "instinct" itself is an admission of ignorance. What follows from instinct would not be a residue because it would have the same status one thousand years later. A residue would have to be considered in the light of social and cultural development. The instinctual eye flutter,[228] for example, would not be considered a residue because there is nothing presumably that it could be a residue of. Even if Pareto's list of residues is suggestive, no account is given by him why one list rather than another should be preferred. Kosambi, for his part, attacks the problem of residues from the point of view of how people make their living. That becomes the key to survivals. Correlations between the instruments and relations of production give us the answers we are seeking. After flint was generally abandoned for knives, flint knives survived because the wounds made by them cannot easily infect. Kosambi points out that the Jews used stone knives for circumcision for this reason even after metal was in general use.

Another example of a survival coming down to contemporary times is the *holi* spring festival which originates in ancient agricultural rites. Today, during this festival people throw colored dyes on each other at one level of celebration. They also dance around a large camp fire at another, or enjoy sexual license and promiscuity at still another. The sexual license was once needed as a stimulus when the diet was poor, life hard, and procreation difficult, says Kosambi. Today the sexual stimulation for procreation has changed to "depravity" because of better food and changed attitudes towards sex. In this instance, Kosambi explains the origin of *holi* but not its survival. He is evidently content to list survivals without explanation. He does, nevertheless, predict that the caste system will disappear because "the brahmin priest is out of place in mechanized life; machines run by scientific laws do not justify a caste hierarchy."[229] It is evidently more efficient to promote machinery than to preach against caste. Perhaps this is not as perceptive as it first appears. The Christian priest and Jewish rabbi have persisted despite the advanced machinery in the United States. It would appear that a more plausible explanation of survivals must be worked out. Kosambi is much stronger in accounting for origins than for residues. This is seen not only in *Ancient India* but also in *Myth and Reality* (1962)[230] which has such original contributions as chaper iii, "At the Crossroads : A Study of Mother-Goddess Cult Sites," and his first chapter entitled "Social and Economic Aspects of the Bhagavad-Gītā." In the analysis of the gītā he makes sense out of the twists, turns, and contradictions to be found in that sacred text. The work of Kosambi leads indirectly into the study of the history of philosophy being conducted by Debiprasad Chattopadhyaya (1918), the founder of developed histodiamat studies of Indian philosophy, an area in which M.N. Roy and S.A. Dange, as laymen, preceded him on a somewhat popular level.

DEBIPRASAD CHATTOPADHYAYA AND SCIENTIFIC PHILOSOPHY

M.N. Roy's own work was of course not historical and dialectical materialism, although he himself claimed to be a materialist. His work tended toward mechanical materialism with elements of bourgeois "pluralistic" humanism because of his strong reaction against revealed religion. Close to Lenin, trusted by Stalin, a mentor of Ho Chi Minh, Roy really believed in the primacy of ideas in the development of human community and hence was never able to overcome his idealism. Yet he was a founder of both the Communist Party of India and the Communist Party of Mexico.

Debiprasad Chattopadhyaya, on the other hand, has been a leading practitioner of histodiamat philosophy for many years and recognized as a world leader in this field both unofficially and officially. A prolific scholar, his main early work is *Lokāyata*, A Study in Ancient Indian Materialism (1959), now in its third edition. Other equally important works have appeared more recently: *Nyāya Philosophy* (with Mrinalkanti Gangopadhyaya) (1967-73) in four volumes, *Indian Atheism* (1969), *What is Living and What is Dead in Indian Philosophy* (1976), *Science and Society in Ancient India* (1977), *Lenin the Philosopher* (1979) and *Studies in the History of Science in India* (1982) in two volumes. These have combined the best scholarship of Britain, France, Germany and the Soviet Union to transcend all other historical work in Indian philosophy in the twentieth century. His work has already been officially recognized among German and Soviet scholars as preeminent and the culmination of the work of his teachers such as Walter Ruben in Berlin. He counts among his admirers and students an array of scholars from Japan to Europe, including the present writer who is constantly in his debt.

Debiprasad conceives the history of materialism to be the history of the philosophy of the people. Indeed, the word, "*lokāyata*" meant "the philosophy of the people." He has been anxious to reconstruct this philosophy of the people in such a fashion as to make it clear to all students of philosophy. In so doing he has been the first Indian to do research and restoration on an extensive scale. George Thomson, the English classical scholar and writer on political economy, earlier did the spadework and summary for ancient Greece in his *Aeschylus and Athens* and *Studies in Ancient Greek Society*, of which Vol. II is entitled *The First Philosophers*, an extension of, and development from, the work of his Cambridge classics teachers including Sir William Ridgeway, F.M. Cornford, and Jane Harrison and not overlooking Gilbert Murray's contribution via Oxford. Debiprasad has also been influenced by Joseph Needham's work on China.

The first step in the Debiprasad reconstruction was to work out the implications of mother-right with its characteristic ideology in India. This is contrasted with its opposite, male domination. These two threads of influence are seen to be of basic and overriding influence in the development of early Indian idealism and materialism. Employing whatever sources invited use, he carefully built up an account of ancient Indian society—superstructure as well as infrastructure. Beginning with the ancient *asura*-view first noted in the *Chandogya Upanisad* by the Indian historian of philosophy, Surendranath Dasgupta (1887-1952) who was his teacher in Calcutta, Debiprasad moved on to tantrism and the complex of archaic ritual practices closely related to the attempt to control agriculture through magic.[231] Emphasis is placed on femaleness and its relationship to procreation and fecundity. Sexual union, for example, was believed to be an aid to *artha* or material well-being as contrasted to *dharma*, which is later associated with abstinence and non-material concerns. *Dharma* appears only with class-society, according to Debiprasad, where it is used for ideological purposes in the same way that the golden mean was used in Greek society to curb "excesses" of consumption.[232] From tantrism, Debiprasad pursues the comparative method to wring out of the Indian scene as much information and insight as he can. He stepped intrepidly into the realm of totemism, group life, magic, religion, the appearance of the early vedic songs, folk, culture, and the emergence of the Upanisads. This Indian development parallels in a number of significant ways the development of the Ionians from totemism to the appearance of tragedy and religious philosophy as depicted by Thomson in *Aeschylus and Athens*. Indian philosophy is thus given its proper socio-historical base, an achievement of the highest order and in richness of detail, deserving of our applause. It can now be said that without a knowledge of Debiprasad Chattopadhyaya's *Lokāyata* it is impossible to have a scientific undertstanding of the development and meaning of Indian philosophy. We should also keep in mind the original impetus given in this direction by Walter Ruben and his teacher, H. Jacobi.

His second major work, in English *What Is Living and What Is Dead in Indian Philosophy* addresses itself to the questions as to the sources and social function of Indian idealism and the refutation of idealism. He begins with a discussion of pre-philosophical songs of the R̥gveda, the fantasy fulfillments of the Brahmaṇas, which is a period in which man no longer struggles against nature only, but against other men. It is during this period that a sacredotal caste develops and changes the poetic forms into the "cumbrous" and "insipid" prose that we find in early Indian literature. As Debiprasad says:

> One reason for this insipidity is the tendency to evolve symbolic interpretations out of ritual trivialities, in the course of which scraps of R̥gvedic verses are often quoted without their context and with strange meanings read into them.[233]

The gods called Maruts are transmogrified into the common people, while despotism is transformed into Indra and Varuna. It says in the Satapatha Brahmana that"Varuna, doubtless, is the nobility, and the Maruts are the people. He (the priest) thus makes the nobility superior to the people. And hence people here serve the Kṣatriya [king, noble], placed above them."[234] The exaltation of mental versus physical labor is comparable to that demonstrated by Plato and his class in ancient Athens. Philosophy becomes a way of looking at the law, and the law is that from which the nobility derives its political power. We think of Plato's *Laws* and compare it to the Bṛhadaranyaka Upaniṣad (i. 4·11-14) where nothing is higher than the Law which is Truth.[235] This gives us our first clue to the social function of Indian idealism.

During this period begins the isolation of philosophical activity from the tools and techniques of the manual workers, through which contempt for physical science is accompanied by an illusion "resulting from the coercion of consciousness to a peculiar process of introversion."[236] What Debiprasad calls the "emancipation of consciousness" occurs after the appearance of early vedic poets, who were obsessed with physical survival. The writers of the Brahamana literature indicate, however, the emergence of a leisure-class engrossed with the problems of "stabilization of the political power of kings and their ideological apologetics." Their main emphasis was upon control through fear, a technique still popular throughout the world. But philosophy does not flower in such an atmosphere. "In the *Brahamanas* we see priests rambling in the graveyard of primitive rituals. . . ."[237] The philosopher first emerges in the Upaniṣads "when the leisure-class fully stabilizes its own power and can afford to have the serenity and tranquillity of unruffled contemlation."[238] But with this Augustine peace comes danger : the danger of separation of thought from production, separation of brain from hand, nature from mythology and ritual. Worse than mere contempt for the physical sciences is "the creation of an illusion, resulting from the coercion of consciousness to a peculiar process of introversion. Knowledge is no longer intended to be the knowledge of objects. It wants to be knowledge of the subject itself—of the bare ego or of the pure self."[239] With this cogitation comes delusions of gradeur, the greatest of which is the belief in the omnipotence of the self or ego. This self then dictates to reality and demands to be treated as the only reality. "The result is lofty contempt for the material world, in which the philosopher himself has his being."[240] We have already noted this contempt in such philosophers as Sri Aurobindo, but there have been thousands more like him in recent times. With this contempt for physical reality, ". . . . the philosopher's consciousness runs the risk of imagning that it can rise to ever higher and ever more remote conditions where only thought remains and the things thought of fade out."[241] We have seen similar occurrences today in Husserl and Heidegger in the West. The Indians have no monopoly on this self-delusion. It occurs to a lesser degree among the formalists and subjectivists—from the followers of Mach and Vaihinger to Carnap and Popper. One could scarcely bring out more clearly just how insubstantial philosophy has become the philosophy of one subjectivizing world; how idealistic philosophy reflects delusions more than reality.

With these delusions develop the cults of secret wisdom—from the Upaniṣadic seers to Wittgenstein. This secret wisdom is immortality in early times. Now it is purely formal knowledge of a subsisting Arcanum. But only the leisure class can hope to understand this secret. The sharing-gods like Bhaga and Amsa are not invoked by the leisure-class philosophers because those two were representative of communal life before Upaniṣadic India. The renowned Yājñavlkya is enabled by means of his "secret wisdom" to a mass considerable property which he wants to settle before his retirement." He is paid by his patrons for initiating them into the "secret wisdom" of immortality. Therefore, he does not have to work for a living. Yājñavalkya is retired on the strength of his alleged wisdom, retired to a life of contemplation. His life of pure contemplation does not fit his working for a living. But he is not as slick as the later idealists who split life into the unreal practical life versus the real contemplative life. He naively admits that he is interested in cattle.

Philosophical parasitism such as Yājñavalkya's is dangerous to the true philosopher. "It kills the philosopher's spirit of interrogating nature, coerces his consciousness to total introversion and makes him a philosopher of pure spirit, for which death is as meaningless as birth." Or it leads him much later in history as it did Heidegger and his followers, to consider death even more important than birth. In any case, if we believe the Upaniṣads, Yājñavalkya ends up with 62,500 cows. Since wealth at that time, was counted in cows as in Rome (*pecunia*) [pecus-cattle] in the early days, Yājñavalkya's patron had made him a rich man. As a rich man he worried about affairs of the spirit. Today such philosophers are neo-Kantian. They have their

Knighthoods, honorary degrees and investment cows.

The parasitism of the priest-philsoophers is referred to by the Buddha in the *Ambattha-sutta* DB ii, 130. But the Buddha "is evidently transforming the reality of the rudimentary control of nature of the ancient Vedic poets into a romantic picture of their great asceticist."[242] At least the Buddha exclaims over the relative asceticism of the ancient poets versus the parasitical ones of later times.

Not only does Debiprasad show the development of parasitical idealism in Hinduism, but traces it in Mahāyāna Buddhism as well. This may be called from Buddha Reform to Buddhalogy. From the Buddha we turn to the many Buddhas and Bodhisattvas. These have little to do with the real, productive, material world. The historical Buddha vanishes to be replaced by supernatural forms. The Zen priest who chops up the Buddha for firewood is a good example of post-Buddhist superstition under which the world becomes illusory. Such latterday Buddhists, Debiprasad says,

> [hope] by refuting the concept of the creator. . . . to prove that the creation of the world—and hence the world as such—is no more than a fiction. This point is later reiterated by Daudapada, whose view is known as *ajanta-vada* or the theory that the world never comes into being.[243]

Debiprasad notes that the Buddhist idealists, the monk-philosophers who push the Mahāyāna claims, "are relieved of the worries of their own material existence" to be able to "evolve new and much more crafty considerations not only in defence of their own outlook, but also in refutation of their opponents, creating thereby for these opponents the need of a better and sounder defence of their own positions."[244] Debiprasad considers their arguments to be both useful and detrimental, depending upon whether one approves of the growth of idealism on the one hand or on the development of arguments, speculation, and critical analysis on the other. Evidently everything has a kind of inevitability about it until mankind puts a stop to it—which is another form of inevitability. There is inevitability without consciousness and inevitability with consciousness. When the latter occurs, then a new kind of inevitability appears.

The social function of idealism is next taken up by Debiprasad. First of all, Debiprasad points out that philosophers and law-givers may often be at loggerheads since law-givers are not interested in truth but in social order. Some philosophers are critical of the laws, but not the extreme idealists. They oblige the law-givers by inventing theological arguments to show that free thinking or uninhibited reasoning is intrinsically fallacious. For this very reason, idealists often get a great deal of political and social boosting and the law-givers "eagerly proclaim that their view represents the quintessence of wisdom." We see this kind of alliance between the pragmatic philosophers and law-givers around the turn of the century in the United States as excuse for individualistic, *laissez-faire*, business enterprise ideology just as the monopolies are being developed. Endowments from business to universities are another example of close ideological affiliation. (Harvard University, where Henry Kissinger taught before advising Nixon, received sixty-three million dollars in gifts in 1979 alone. Gifts to Columbia University rose markedly after Zbigniew Brzezinski served with President Carter.)

Of particular significance to the student of Indian philosophy is Debiprasad's analysis of the philosophy of Sāṁkara which is based, he maintains on class contempt of the *śudra*, the *śudra*, who by nature is "debarred from the philosophical wisdom he speaks of, the clear corollary is that the question of the *śudra* ever attaining real freedom does not arise."[245] His job is back-breaking toil with the hope of eternal bondage. The reason for thus being cast into the outer darkness is that first of all, the *śudras* have not studied the vedas, and second, they have not partaken of the *upanayana* ceremony through which they can receive the status of *dvija* or upper class (in a world composed of *dvijas* and *śudras*). One explanation for this hard stand is that the law-givers who have laid this down are infallible. Then Debiprasad quotes the legal authorities of India putting the *śudra* forever in contemptible bondage. Of this he comments,

> For sheer barbarity, even the raw defence of slavery in ancient Greece and Rome can hardly vie with the authorities quoted by Samkara. It is therefore surprising that some of the modern admirers of Samkara are fond of writing on the humanism of his philosophy, when the philosopher himself quotes with enthusiasm all these inhuman legal measures against those whom he knows as the toiling millions of the country [such as pouring molten lead into the ears of unqualified hearers of the vedas, or slitting the

tongue of someone who is not fit to pronounce the vedas and so forth).[246]

Sāṁkara is more reactionary than Plato. He is proud to be the tool of the law-givers. Sāṁkara is an apologist par excellence, working overtime for the rulers. By his social views we know him to be against Indian progress, a mythmaker, a philosopher who "shows his servile admiration" for the status quo order of the law-giver. It is only with the Lokāyatikas that we find resistance to counterproductive idealizing. These philosophers see through the apologetics of the law-givers.

Other Indian philosophers also fight against the illusions of idealism. They include, says Debiprasad: Kapila, exponent of recognizing primeval matter as the first cause of the world, Kaṇāda the so-called atom-eater, Gotama the philosopher with "eyes on his feet", Srilabha the Sautrantika Buddhist, an important target of the idealistic Mahāyānists. Subhagupta the Baibhaṣika, Akalanka the Jain logician and atomist. Prabhakara and Kumarila of Mīmāmsā philosophy who try to refute idealism, Uddyotakara, Jayant Bhatta, Vacaspati Misra, Udayana—all defenders of atomism and logic from the standpoint of Nyāya. Last of this great tradition is Gadadhara of Bengal, of the seventeenth century. "The antithesis of idealism in Indian philosophy is..at best a descriptive epithet for an amorphous mass of heterogeneous disapproval of the idealistic outlook. It is futile to expect of it a coherent alternative to idealism."[247]

It seems clear that without a knowledge of Debiprasad's *Lokāyata* we cannot understand the main ideas and arguments of classical to modern Indian philosophy and without *What is Living and What is Dead in Indian Philosophy* we cannot fathom the development of Indian metaphysics, epistemology, and ethics as we find them today.[248] As a worthy student of Surendranath Dasgupta, Debiprasad has become his successor as the most important historian of Indian philosophy of his own time.

ABDUR RAHMAN: THE CONFLUENCE OF SOCIAL AND PHYSICAL SCIENCE

If one studies science in any given period, Rahman maintains, one finds two simultaneous processes in operation. The first is the extension of a theoretical outlook through its applicability to new areas. The second process involves the rejection of an already accepted outlook. Early empiricism was partially a reaction against "over-philosophizing".[249] Today a reverse process is taking place. Fragmentation of knowledge is leading to a modern trend of integration of information. To understand his expansion on this theme, see his *Science, Technology and Economic Development*, (Delhi : National, 1974), esp. pp. 198, 214. Also see his "The Psychopathology of Committees", pp. 25-32. Consonant with this movement is the value of Engels' contribution to the history and philosophy of science since "..he places science in its historical perspective," according to Rahman.[250] This is tantamount to saying that Engels treats science in its interconnections instead of an isolated form. In what ways Engels' thought is useful today raises important questions in both science and the philosophy of science. It should be noted, first of all, that Engels wrote at a time when evolution created a revolution in biology. Today we are in another revolutionary period in which biology is being prodded forward by genetics and electronics.[251] It is useful, Rahman thinks, to compare these two periods to see what we can learn from considering them together.

According to Engels, science is a social process dominated by the general outlook prevalent in a society. By comparing the perspective of classical Greece with that of his own day at the closing of the nineteenth century, Engels attempted to clarify the prevalent outlook of each. Whereas the pre-Socratics believed that the world had emerged out of chaos and developed in a certain way into the reality we know, the scientists of that period, for the most part, were dominated by theological considerations. This tenor continued through the medieval period in Europe. We find theological considerations to be sometimes uppermost in Aristotle as interpreted by Thomas Aquinas, although Albertus Magnus, Roger Bacon and others heralded the new outlook.

The developmental view came into being as may be seen in the historicity of Kant's thinking in astronomy, which was picked up by Laplace and Herschel. Lyell made the same kind of sense out of geology as Kant out of historical astronomy, showing the gradual effects of a slow transformation. To overcome the older, static, thinking is by no means easy, according to Engels, for indeed: "..tradition is a power not only in the Catholic

Church but also in natural science." And one may add the parallel thought that static thinking in Indian philosophy occurs not only in Hinduism (and Islam) but also in the idealistic tradition so much loved by the philosophers themselves.

Engels puts his finger on a major difficulty that was faced not only in his day, but is still a problem today. Rahman quotes him as saying:

> Empirical natural science has accumulated such a tremendous mass of positive material for knowledge that the necessity of classifying it in each separate field of investigation systematically and in accordance with its inner interconnections has become absolutely imperative. It is becoming equally imperative to bring the individual spheres of knowledge into connection with one another. In doing so, however, *nature science enters the field of theory and here methods of empiricism will not work*, here only theoretical thinking can be of assistance.[252]

Some of the attempts to modify empiricism are apparent in field and gestalt theory in psychology (even in some of Jung's abortive attempts to give historical meaning to psychological states although it is doomed because of Jung's erroneous idealistic presuppositions), in systems analysis, which is a belated recognition of dialectics, incipient in the work of Thomas Kuhn and Paul Feyerabend, both of whom try to break out of mindless empiricism. They are finding it difficult to succeed, however, because of overemphasis upon the formal rather than the social components of science. Bearing on this, Rahman quotes Engels to the effect that :

> In every epoch, and therefore also in our, theoretical thought is a [sic] historical product, which at different times assumes very different forms and, therewith, very different contents. The science of thought is, therefore, like every other, an historical science, the science of the historical development of human thought. And this is of importance also for the practical application of thought in empirical fields. Because in the first "eternal truth" established once and for all..[253]

It is useful to note that a number of logicians in several socialist countries (USSR and DDR) went so far in taking the dictum of Engels to heart that they began to attack traditional, formal logic as being superseded by dialectical logic.[254] More recently, there has been a reaction against this one-sidedness so that it is now generally considered that each logic is relatively useful and true in its own domain.[255]

Scientists and philosophers of science have a great deal to learn from Engels, says Rahman, even though he has been dead for almost a century. India's legacy of colonialism left a situation where modern science was divorced from national thought, especially separated from ancient and medieval Indian scientific tradition. Consequently, Rahman maintains, "..the thought of our antiquity, in spite of its materialist content, has only its idealistic exponents. Science, being divorced from the national-historical traditions, looks to Europe for its theoretical guidance.."[256] Rahman might have added that this is generally true not only in science, but also in philosophy. The late Walter Ruben, former Director of the Institute of Indology at Humboldt University in Berlin (formerly Berlin University), chided me recently for overstating the European influence on Indian thinkers to the detriment of their indigenous claims. Perhaps Rahman shares my predilections in this respect. Ruben, I believe, never had as many face-to-face relations with Indian philosophers and scientists as Rahman certainly, and perhaps was more isolated than American Indologists over the past thirty years. Since the end of the second World War we in North America have been closely associated in the West with dozens of Indian philosophers in a constant stream. See Ruben and Hiltrud Rüstau, *Mainstream*, Vol, XIX, No. 20 (January 17, 1981), pp. 32-34.

Furthermore, India has not only imbibed colonial traditions in science, "..but due to the historical factors of its birth, considers those features as 'true' features of science."[257] We have already seen how this mistaken generalization led many of the fighters for Independence to be anti-scientific [M.N. Roy] because they thought that science was essentially emperialistic and European in nature. Indian scientists themselves [as well as the philosophers of science] are victims of common superstitions and pseudo-religious traditions which can be overcome partly when these workers begin to understand their own history and its traditions. Therefore, such workers should read not only works like Engels' *Dialectics of Nature* but histories of Indian science like those of Rahman himself, and cultural histories like those of Kosambi and Debiprasad. There are scores of others to be dug out of libraries and progressive bookstores because of the present period of remarkably burgeoning Indian historical studies.

Rahman's insistence upon a serious reappraisal of attitudes toward science in India makes him a pioneer in heightening the consciousness of the scientific community and the community surrounding scientific work.

DILIP BOSE AND THE DIALECTICAL APPROACH TO SCIENCE

Another scholar working in the area of philosophy of science, and like Rahman not a philosopher by profession, is Dilip Bose who has written the essay: "Engels, Dialectical Materialism and Science".[258] Bose cuts across the physical, biological, and social sciences in this book, but not in a spurious way, as in the so-called "unity of science" movement which is formalistic and one-sided. Unlike the logical positivists and logical empiricists who turned science into a formalistic or subjective undertaking, Dilip Bose approaches science from a profound historical viewpoint that will not evaporate into agnosticism. The neo-positivist (idealistic) approach *at its best* is well-summarized by Ernest Nagel, who says:

> The long history of science and philosophy is in large measure the history of progressive emancipation of men's minds from the theory of self-evident truths, and from the postulate of complete certainty as the mark of scientific knowledge. Some of the major turning-points in that history consist in radically diminishing the class of statements certifiable simply by a rational insight into their truth.[259]

Although Nagel thinks that his approach is historical because he refers to history, it is indeed abstract and ahistorical. One can fill in the blanks by already knowing the major turning-points in the development of science as understood by histodiamat or its closest complements which includes the biological and social vectors. Nagel's viewpoint usually hinges around the formalistic, rationalistic considerations having nothing to do (so far as we can tell from his account) with man's productive existence or the technology that was developed to create the changes that led to scientific "revolutions". "Self-evident truths" are not nearly so important as truths that keep afloat through traditional lifebelts. "Rational insight into their truths" is simply a mystification of the complex process of ideology. Nagel is abstract and ahistorical in speaking of "progressive emancipation". When Nagel turns to probability theory itself he does explain certain productive and material features, including the rise of commercial insurance in Babylon, Greece, and Rome: rates on bottomry and marine insurance, annuity problems in expanding Holland and England, games of chance in the Renaissance and the seventeenth century France developing into the "statistical view of nature." These considerations are very helpful.

Nevertheless, Nagel usually does not get to the bottom of the social events that brought about significant change. He stops far short of having a philosophy of history, without which there can be no philosophy of science worthy of serious consideration. There are moments, however, when he appears close to putting forward a kind of dialectical materialistic philosophy, but then draws back like a crab in his shell. In this feinting towards a straight-forward, ontologically-grounded dialectics, he reminds us of Dewey who with praxis neared the threshold of histodiamat. Dewey, the teacher of Nagel, wrote in his *Theory of Evaluation* (1948):

> We are living in a period in which emotional loyalties and attachments are centered on objects that no longer command that intellectual loyalty which has the sanction of the methods which attain *valid* conclusions [note the use of "valid" rather than "true" —my italics—DR] in scientific enquiry, while ideas that have their origin in the rationale of inquiry have not as yet succeeded in acquiring the force that only emotional ardor provides [one has the "feeling" that Dewey is about to say something momentous]. . The split which exists in present social life between ideas and emotions, especially ideas that have scientific warrant and uncontrolled emotions, that dominate practice, the split between the affectional and the cognitive, is probably one of the chief sources of the maladjustments and unendurable strains from which the world is suffering [here Dewey is psychologizing-away a problem that is based on a class split]. . The *practical* problem that has to be faced is the establishment of cultural conditions that will support the kinds of behavior in which emotions and ideas, desires and appraisals, are integrated.[260]

Yes, assuming that we know what "intellectual loyalty" is, we couldn't agree more with Dewey. Now, how do we bring about these cultural conditions conducive to the scientific outlook that both Dewey and we wish? Dewey and Nagel are silent on this question because they do not themselves have a *political philoso-*

phy through which to integrate their scientific philosophy with concrete conditions.[260] Cultural conditions are not merely the affect of the split and maladjustment, but are the cause of them. They include ideological considerations. Even so, this is one of the better and concerned statements made by an instrumentalist who has more of an attitude than he has a philosophy.

Because there are Deweys in India as well as in China (Dewey's student Hu Shih, a supporter of gradualism and ultimately Chiang Kai-shek, was a well-known philosopher in Taiwan in his later years and the most visible Chinese anti-Maoist philosopher until his death in 1962. His solution to the overarching problems of China was "Save the nation by study" and to realize that China's characteristics were "femininity, cosmopolitanism and passivity." Hou Wai-lu, one of his harshest critics, claimed that Hu Shih tried to weld together Chinese feudalism and American imperialism. [By so doing this would link theory with practice]. Let me restate and compress Dewey's passage in language that can be immediately understood. The first statement of the quotation I take to mean :

We are loyal to beliefs and institutions that do not command the respect of the sophisticated scientific community because they are not grounded in objectivity. The second statement I take to mean : a chief source of social (and individual) maladjustment is that our ideas are out of phase with our emotions. The third statement I take to mean : we ought to bring about a society in which emotions and ideas are integrated. This is to say that we should integrate ourselves by adopting scientific presuppositions and procedures and/or we ought to form a new social system based on scientific procedures.

That brings up the next question which is: why are we loyal to unscientific beliefs and institutions? Dewey cannot, dare not, will not, answer, although he hinted at the answer. If he should, he would be led to the unmentionable examination of class structure in our society.

Scientists naturally have dual feelings when they consider the consequences of their work. Many of them, however, seem quite oblivious to social or humane aspects of science. They consider themselves bondsmen of whomever coughs up the research money. They are the scientific slaves of the twentieth century and are perhaps replaceable by robots.

A great deal of the conflict between so-called emotion and so-called ideas (Spinoza and Hegel knew that they were inseparable anyway whereas Dewey is a reluctant Christian dualist) because of the injustices rampant in class society. For what purposes should science be used? Should it be used for increased military buildup, space stations for Star Wars, environmental pollution primarily where the working class lives, football parades, selling tires, TV commercials selling dogfood and preparations that are unneeded by those eating prunes and other roughage, visual (fictionalized) education between commercials as how to kill your grandmother (how someone else did it), getting rid of your parents and making life miserable for your employees? The political-economic basis of the injustice in a class society must be revealed—that is part of the work of social science. Social science has an obligation to also reveal the injustice in an allegedly classless society.

Dewey must have had some intimation that institutionalized science would never be free from dual loyalties unless such science was practiced in a society which was free from domination by a calloused clique representing only its own desires. Representative government is a shallow hypocrisy when corporations spend hundreds of millions of dollars getting "their man or woman" elected. I discussed that with an English economist now with the World Bank as we took tea on the veranda of a residential hotel in Calcutta. He thought that buying elections was admirable. He believed that not only the Americans and Japanese but also the English should do more of it. I was dismayed especially since he was the son of a gifted English philosopher.

This pragmatism and instrumentalism is expected by post-World War politicians whose main concern is the Cold War. But Dewey's approach is surprising. It imbibes the idealism that removes the material base and its relations and leaves the question floating on the psychological surface. As George Novack has said: "..instrumentalism..feels most at home in a twilight world where sharp distinctions are blurred, outlines are vague, and the precise places of things are uncertain.. [feeding on] ambiguities and patchwork."[261]

As the generally accepted intellectual leader of progressive forces that attempted to assert the claims of the middle-class (intellectuals, small business-people and technicians) as opposed to the predatory corporate class, Dewey believed that the Declaration of Independence, the Constitution and the Bill of Rights were sufficient to serve as a basis for a new hope for democratic institutions. Yet this bourgeois democracy in which Dewey believed, changed its stripes from mixed progressivism and conservatism on the American science to virtual apologetics upon the luxuriant flowering of monopoly capitalism. From apologetics it turned to craven acceptance under Carter and Reagan and the great corporation celebration and the crushing of the working class who were to pay for the world-wide capitalist depression. Earlier these instrumentalists had caved in (according to Randolph Bourne in 1917) to the military adventurists and then groveled before the Truman-McCarthy witch hunters.

Dewey's social indeterminateness and personal blindness of actual industrial conditions helped propel instrumentalism into open collaboration with the new social reaction beginning in the second phase of the Great Cold War opened by Truman with the encouragement of Winston Churchill in 1946. Liberal sympathy for even tepid socialism evaporated along with the Stalin disclosures and the ranks closed in defense of God, Christianity and Judaism, free enterprise and the purchase of elections on the open market.

One sign of the times was that the foremost follower of John Dewey who had once advocated Korschian socialism with a Marxist tinge defended the practices and aims of Richard M. Nixon. Dewey's beating around the bush for seventy years led him on the one hand to deplore the lack of acceptance of science (and attack the irrationalism of religion) and on the other hand, made him keep it a secret why social science cannot be accepted by the dominant class under capitalism. Social science, as Bernal pointed out, is not neglected because it is so difficult and complicated—there are so many variables and so much more emotional attachment—but because the ruling clique forbid its significant use by a hundred subterfuges, including foundational grants for trivialities.

The great housing (low-cost) projects I saw in East Berlin in 1978 would be quite impossible under capitalism. More to American capitalist taste are projects in which condominiums with two bedrooms cost $200,000 and up in Manhattan. Richard Nixon's New York apartment sold for $2,450,000. I once visited a friend in his father's apartment in Chicago with six bedrooms which cost $850,000 twenty-five years ago. In the large cities of the United States "improvement" of central cities was called even by social scientists "negro or black removal" because their housing was taken over by hotel entrepreneurs and civic centers which made it possible for out of town salesmen to be more comfortable in the central city. No substitute housing was offered the citizens who were forced from their dwelling places. We heard no outcry from architects who said that they would not work on such schemes.

With this scarcely elevating digression, we turn again to the view of Bose who regards Engels' *Dialectics of Nature* as the most important book in the philosophy of science in the nineteenth century. In an important sense one would not be exaggerating to say that it was perhaps the only serious work in the field at that time. The *Dialectics of Nature* stated that matter and motion are inseparable, that there are qualitatively different forms of motion which require different sciences and there is a dialectical transition from one form of motion to another. The work attempted to bring out the internal connections among the sciences. Engels, by the 1890's, in his seventies, was never able to complete this work because he spent the last twelve years of his life editing volumes II and III of Marx's *Capital*, from the copious notes left by Marx. Engels believed that editing *Capital* was more important than doing his own research on the philosophy of science.

Of great interest is the statement by Engels concerning models in the letter he wrote about the theory of evolution which he believed undervalued co-operation as opposed to the struggle for existence. He said: "The Darwinist teaching of the struggle of existence is simply a transference from society to living nature of Hobbes' doctrine of *bellum omnium contra omnes* [the war of all against all]..and of the bourgeois-economic doctrine of competition together with Malthus' theory of population."

Bose might have recalled that Aristotle's *Metaphysics* was heavily dependent upon the model of Greek social institutions as was his treatment of women in the *Poetics* or the modelling of the universe in Confucianism and Neo-Cofucianism on the social institutions of the Chou and Shang Dynasties.

According to Bose, the view of Engels that:

> ..these antagonisms and distinctions thought to be found in nature, are only of relative validity..that on the other hand their imagined rigidity and absolute validity have been introduced into nature only by our reflective minds—this recognition is the kernel of the dialectical conception of nature..

The ultimate goal that Bose anticipates is a view of the philosophy of science that breaks out of the pattern of current Western bourgeois piecemeal examination. When this is achieved, it will be possible to bring together the conceptions of science in one system as J.D. Bernal has attempted in his *Science in History*.

Bose has made a good beginning of the reform of philosophy of science by bringing out the importance of the dialectical approach to its understanding and development. Not only Engels' and J.B.S. Haldane's work must be carefully studied, but also that of Bernal, Joseph Needham, Lancelot Hogben, Hyman Levy and other dialecticians working in the philosophy and history of science.

SUMMARY AND CONCLUSION

For the most part, Indian philosophy of science is idealistic and subjective. It uses Western European and American paradigms which are for the most part positivistic and idealistic, hence static, rigid, metaphysical (in the non-Collingwood sense of dealing with presuppositions), and non-dialectical. There is in addition to these qualities, the inherited attachment to classical Indian conceptions such as supermentality, spiritual evolutionism and a defensive attitude regarding religion. Beyond these difficulties, some of which Western philosophy of science shares, are the more universal among bourgeois thinkers. Some of these are the separation of theory and practice (Popper), fact and value, hand and brain as well as a smug disregard of historical conditions, philosophy of history, social sciences outside of logic and statistics and dialectical philosophy. One economist I have known with a doctorate from Harvard University, chairman of a large economics department in a major university laughed until the tears came to his eyes when I suggested that perhaps Marx still had something to say at the end of the twentieth century. "Why all that went out in the nineteenth century" he said as he squeezed my elbow.

One advantage that subjectivist and idealistic Indian philosophers of science yet show over their Western counterparts is the possibility of shifting loyalties to such theoretical views as would take into account the ethical and developmental considerations in philosophy of science. India is developing whereas most of the Western capitalist countries are contracting. Their overall increases are either largely inflationary or fluff. They may yet in India emphasize the social dimension of their outlook simply because the contradictions in Indian society may be even greater than in many a Western society, although both Eastern and Western capitalism are in enormous trouble as we know on the basis of considering national debt and interest settlements alone. There is a stronger drive in India towards socialist solutions and accommodations which must have an effect on even the most ivory tower bound Indian spiritualist and subjectivist. And the members of the present Indian government are models of flexibility when compared with what has been installed in the White House for the past four decades. The crumbling and demoralization is surely as portentous as that admired in the last days of the Ch'ing Dynasty.

Nevertheless there is room for considerable hope for the future. As we have seen from the accounts of the realistic and materialistic philosophers of science in the latter part of our chapter, there are vigorous and progressive minds at work trying to bring the real world and the philosophy of science together. These valiant thinkers, swimming against the subjectivist and reactionary current have caught the vision of science as a unified endeavor to bring incomparable advantage to the mass of mankind. We must remember also that as is true of sections of the British ruling class, the Indian ruling class encourages anti-scientific attitudes and institutions quite as much if not more than the opposite.

It is positivism and neo-Kantianism with their admixture of revivalism which covertly or unconsciously try to keep science in the hands of the oligarchs for their personal profit (especially in superarmaments) and luxury when theyare notdenouncing science to curry favor with the old reactionary community. The new philosophy of science developing in India (for the most part outside the philosophy departments) is an extension of the desire for development and freedom of knowledge for the whole population, especially the working class who supports the whole edifice with its energy, sweat and patience.

III

CONTEMPORARY WESTERN SUBJECTIVISM IN THE PHILOSOPHY OF SCIENCE AS A BACKGROUND TO INDIAN PHILOSOPHY OF SCIENCE

INTRODUCTION

According to the Patriarch Aśvaghoṣa (second century):

> As from the True Reality man knows that there is no objective world. . All phenomena are originally in the mind and have really no outward form; therefore, as there is not form, it is an error to think that anything is there. . . . Without mind, then, there is practically no objective existence. . . . all existence is like a reflection in a mirror, without substances, only a phentom of the mind.[1]

A major basis for subjectivism is clearly demonstrated here by the famous Buddhist sage. It may seem a great distance in time and space from Aśvaghoṣa in the second century in theory and practice to some of our present-day philosophers of science, but they are in truth much closer than one might at first anticipate. Subjectivism has its two main sources in idealism (primacy or exclusive existence of mind) or in epistemological agnosticism (human inability to know whether he or she can know reality). Whereas many of the Northern Buddhists accepted the idealism of Aśvaghoṣa, today a large segment of philosophers of science in the West stem from the agnosticism of Hume or Kant. Kant's agnosticism was partially inspired by the needs of fideism, while Mach's agnosticism was enlivened by rapid development in physics which tended to make the more tender-minded physicists and philosophers see an upsurge of immaterialism. According to James K. Feibleman: "We are [still] going through a period in the philosophy of science in which it has been dominated by the theory of the dematerialization of matter."[2] Such theory has been maintained, Feibleman states, from Eddington to Norwood Hanson, but of course we know that it reaches back to at least Mach (1838-1916) or Hume (1712-1776). It was Mach who not only said that a "thing" is a complex of sensations (congeries of sensation for Berkeley), but that causality, necessity, and substance must be rejected because they are not "given" in experience. This was picked up by the logical positivists, logical empiricists and their friends. Western idealistic philosophers of science have developed out of the Humian, Kantian and Machian positivist position and aided by misinterpreted mathematical physics.

The influence of mathematics, logic and formalism has been nearly as strong as the sway of physics. Besides the direction given Western philosophy of science by modern physics, there has been the hold of such autonomists as Peirce, Russell, Boole, Peano, Frege and Hilbert, who thought that mathematics developed according to logical principles quite independent of nature. Hilbert, says Dirk Struik of M.I.T., "......saw the real content in the abstract form."[3] Also influential on the idealistic philosophers of science were Hahn, Carnap, Reichenbach and other logical positivists and logical empiricists, sometimes also referred to as empirio-critics. These first and second generations of philosophers of science following upon the heels of Mach were also sometimes influenced by a third group, called mathematical intuitionists, such as Kronecker, Poincaré, Brouwer, Heyting and Weyl.[4] Whereas the positivists treated logic and mathematics as having nothing to do with material objects of the world (when they were allegedly speaking about objects they were speaking about concepts, words or symbols) the intuitionists believed that mathematics and logic were "constructed by intuition." A major emphasis, then, of the idealistic philosophers of science making up (*double entendre* intended) perhaps 90 percent of all philosophy of science in Western Europe and North America, which dramatically affected other branches of philosophy and social science, was the attempt to construct mathematics and logic within a self-contained shell without objective reference to a world of material objects and processes—which is to say, the world of reality.

Within the social sciences themselves, at least two threads allied to mathematico-logical and quantitative experimentalism emphasis may be discerned. These include the work done by Vilfredo Pareto (1848-1923) who emphasized the logico-experimental" method (where the "experimental" might prove to be "controlled observation." Pareto says:

> A logico-experimental study merely relates facts with facts. If that

> is done directly, merely describing facts that are observable simultaneously, we get pure empiricism [sometimes less flatteringly called "mindless empiricism"]. Empiricism may serve to discover uniformities if, by observation and experiment, one succeeds in distinguishing not more than two categories of facts that stand correlation.[5]

If by this technique fails to find uniformities (rather than "causes", then we may resort, Pareto says, to "hypothetic abstractions".

Besides Pareto and his followers, within sociology is the neo-positivism-cum-pragmatism of George A. Lundberg (1895-1966). Lundberg differs from other positivists in emphasizing biological models instead of physical ones (the general tack of Deweyites). His pragmatic presuppositions tie him up with Dewey's theory of valuation (Chicago, 1939) and Bridgman's operationalism in epistemology. He is concerned with "the most *useful* definitions of such phenomena," with operational definitions of phenomena under consideration. His typically fragmented view of "useful," places his work within mini-operationalism. He does not consider the role of the infrastructure or superstructure and their historical roots. Reality, insofar as it is tangentially referred to, is treated in such a way as to have Lundberg fail to take into account *who* it is that is deciding what is useful and what is not useful. One may grant that this is an attempt to remove the personal equation, yet it does violence to actuality since human ideology plays such a vital part in decision-making at every level. Humans operating in the social sciences are quite different from atoms and molecules playing their roles without discernible ideological-social bearings. There is appeal to experiment but none to history. The caution demonstrated (if it is rather then apriori dismissal of other factors) is laudable from a positivist viewpoint, but Lundberg suggests that social scientists should be "non-moral" and be ready to serve any political regime that is in power. In other words, on his view:

> The services of *real* social scientists would be as indispensable to Fascists as to Communists and Democrats [note Cold War differentiation], just as are the services of physicists and physicians [i.e. social scientists] are the passive tools of interest groups rather than active policy-makers because of Lundberg's metaphysical view of knowledge as basically contemplative.(See *The Behavioural Sciences* : *Essays in Honor* of George A. Lundberg, ed. Alfred de Grazia, Rollo Handy et al (Great Barrington, 1968).

Although Thomas Kuhn might not subscribe wholly to this outlook, his own disregard of the crucial social factors of structure and ideology would not prima facie put him at loggerheads with it either. Lundberg defines value operationally as that toward which people behave so as to retain or increase their possession of it. Value for him would have no essential characteristic that would make it quantifiable, a necessary desideratum in his system. Institutions, Lundberg claims, are mechanisms which men have established to secure their ends [which men, whose ends ?]. His view is subjective as is the view of the Austrian Schools with their economists and psychologists. But he warns us, in agreement with Max Weber, that science "as such" cannot or should not formulate value judgements. Moral and scientific statements are irreducible to each other.

What Pareto and Lundberg share is their concern with quantitative aspects of the world while espousing idealism. They appear to agree that scientists have fought their biggest battles, as Pareto puts it, "against the methods of introspection, etymology, analysis of verbal expression." Bernal sees the heaviest seas surrounding a determined effort by powerful cliques to impress their own ideology (superstitions etc.) on given societies which makes the experimental method difficult if not impossible to employ. But nevertheless Bernal believed in the 1950's that "science" had become in some sense ideologically neutral as a force of production. He might have pointed out that whatever was as a matter of fact neutral could be called science. When Einstein said that "physics is metaphysics" he may well have made a most profound statement in the philosophy of science in this century. That could well have meant that despite ideology, irrationalism, illogicality, traditional beliefs, science did go ahead. In addition to that, perhaps Einstein meant to add that the problems will continue just as they do even in the postulated perfect marriage.

Pareto warned us that:

> Belief that the facts of the universe and their relationships could be discovered by introspection was general in a day gone by, and it still remains the foundation of metaphysics, which seeks a criterion of truth outside experience.

Let us warn Pareto that if one seeks the criterion of truth *within* experience he does not know whether that experience is subjective or objective in its reference. This is the same experimentalism that R.W. Sellars complains about in Dewey. We cannot discover the truth by merely referring to experience, since part of our experience includes the introspective. The use of the deductive or inductive methods may be partly determined by other social factors such as ideology and history. It was Spinoza who solved the problem of Cartesian dualism, not by experiment but by introspection and doubtless deduction and induction as well. He did this despite the warnings of the physical scientists of his day. For him, it seemed quite sensible to say, that in the mind-body problem you don't really have two totally different things but merely two modes of the same thing. That took thought—introspection. Einstein used introspection to see that the hitherto elements of physics and mathematics seen as radically separate were really quite harmonious. No amount of deduction will change the viewpoint. What is required is seeing from a new aspect the same old elements. Not everyone who cries "Lord Lord" or "Experiment Experiment" shall be enlightened. Sometimes it appears that the more determined the social scientist is to be hard-headed in his method, the further from objective reality he seems to wander. The ancient Taoists often have something apposite to say on difficult topics. In this instance one thinks of the utterance of the *Tao Te Ching*.

> Thus rigidity and strength are concomitants of death, but softness and gentleness are companions of life. XXVI (tr. W.G. Old (1904).
>
> The good walker makes no dust after him XXVII (*ibid*).

Sometimes naive realism and introspection will take us just as far as tight-clenched experimentalism.

About fifteen years ago, my secretary proved to me the value of insight. I was attempting to counteract the strong doorcheck that prevented my office door from standing open without a heavy chair being propped against it. My attractive secretary looked at it for a moment and found a small piece of wood which she placed under the door and pulled it sharply towards herself. That took care of that problem. In the meantime I had asked the opinion of several professors as to how to keep the door open without a bulky object standing in the doorway, including a man who had two doctorates: one in physics and one in philosophy. He asked: "Have you thought of nailing the door open?"

Insight is made up, of course, of insight and outsight. There are a mixture of ingredients that include information from experiment, from deduction, from extrapolation, from imagination, and from dialectics of analogy, correlation and interconnections of simple to complex mediation and intercession. Some people have noted that there is considerable overlap in outlook and result between the instrumentalist view of knowledge and histodiamat. It seems to me that, although everything valuable in instrumentalism contextualism, experimentalism, creative cognition, validation of common sense is to be already found in histodiamat long before Dewey's writing, what is missing from instrumentalism is a clear-cut view of the independence of the object from knowing (seen by the critical realist, Roy Wood Sellars, who also has called himself a "reformed materialist"—presumably so he is not mistaken for a mechanical materialist). We have earlier mentioned this difficulty in Dewey. According to Thomas English Hill: "Dewey has insisted that the object of knowledge is something that does not exist prior to inquiry but is a product of the resolution of the unsatisfactory situation leading to inquiry.[6] The object for Dewey, Hill says,..comes at the end the of process rather than at the beginning." This aspect of Dewey was criticized by Arthur Murphy in the Schilpp volume on Dewey and replied to by Dewey.Hill appears to think that Dewey has answered the charge in his reply to Murphy, but I cannot see that he has. To discuss the whole argument here, however, would take us back through the history of epistemology. On the basis of his answer and his other writing, in my opinion, Dewey seems to be incapable of even thinking of an object independent of a knower.

But to continue with the relationship of subjectivism and idealistic mathematics and mathematical physics: we may begin as early as with W. Whewell's *Philosophy of the Inductive Sciences* (London, 1837). Whewell maintained the apriori nature of mathematics giving idealism a much unneeded idealist-subjectivist boost. The reason for the preeminence of mathematical physics in recent times was influenced by its tremendous acceleration in research and invention between 1800 and 1900 on the one hand, and on the idealistic ideology of its practitioners and popularizers on the other.

With the influence of Hegelianism, however, especially in Germany where physics and chemistry were expanding rapidly, a gulf developed between science and philosophy, largely because Hegel's system of nature appeared to experimental scientists to be "crazy". Helmholtz, who sometimes deplored the attitude that philosophy had nothing to say to science, recommended that philosophy confine itself to epistemology. This situation of isolation changed somewhat in 1883 when Mach pointed out the philosophical basis of mechanics. But this "basis" reinstituted Berkeleyanism as seen, for example, in Mach's definition of body: "A body, "Mach said," is a relatively constant sum of touch and sight sensation." In reflecting on this definition one is reminded of Aristotle's statement in *Metaphysics gamma* where he says: "..an accident is not an accident of an accident." A sensation, 'likewise, is not the sensation of a sensation, as Mach's view implies.

Since Riemann had imagined different kinds of space, then space might be considered as a concept derived from imagination. So now we had matter made up of sensations—of what? And space was the result of imagination but not inherent in the human as Kant believed. Mach gets the credit, however, whatever his conscious intention, for a major assault on realism and materialism at the end of the nineteenth century. From Mach to the first edition of Engel's *Dialectics of Nature* in 1927 was a period that saw philosophy of science, such as it was, dominated by idealistic philosophers and idealistic scientists with such brilliant exceptions as Lenin in his celebrated refutation of Mach entitled *Materialism and Empirio-Criticism* (1908). The neo-Kantian Lange, Dampier and others seem to have been unaware that ideology is involved in science and the philosophy of science as well as involved in political economy and ethics. According to J.B.S. Haldane:

> One reason why Engels was such a great writer is that he was probably the most widely educated man of his day. Not only had he a profound knowledge of economics and history, but he knew enough to discuss the meaning of an obscure Latin phrase concerning Roman marriage law, or the processes taking place when a piece of impure zinc was dipped into sulphuric acid. And he contrived to accumulate this immense knowledge, not by leading a life of cloistered learning, but while playing an active part in politics, running a business, and even fox-hunting.[7]

Engels was perfectly aware that writers on the history and philosophy of science were writing from either a materialist or an idealist point of view, with a heavy preponderance of the latter. It was not until J.D. Bernal in *Science in History* that a truly encyclopedic history and philosophy of science appeared which included not only the physical but also the biological and social sciences in the interrelations with each other. His account of social science, although striking many of the most important conceptions, needs to be greatly supplemented and updated.

Phillip Frank is perhaps typical of his school of philosophy of science, which is to say, largely oblivious to the social significance of philosophical interpretations of science. Even so, he is more aware than most of his colleagues since he had at least been threatened by the holocaust. He says in his *Modern Science and its Philosophy* (1951).

> The question whether there really is a matter (a thing-in-itself) that is different from the sensations has no meaning for science, since no experiment can be performed that might possibly settle the question. Every proposition about a gross material thing can also be expressed about sensations. Thus the whole of mechanistic physics can be translated into the language of Mach.[8]

It is easy enough to imagine experiments to refute Frank. The first is to try to sell a man the sensations of an automobile without also selling him the automobile which provides the sensations. In 1969 a salesman who had sold me a Chrysler station wagon handed me the keys to the car. I unlocked it and got in. Immediately I looked for the lever which made automatic long distance driving a possibility—called "cruise-control" but couldn't find it. The salesman motioned me outside of the car and pointed to the sticker listing the standard and optional equipment pasted on the windshield. "There it is, right there."

"But that isn't it. That is simply a visual sensation of its description." I complained.

My wife's aunt complained to us that her amputated leg pained her a good deal. She was having a sensation or a series of sensations of an object that did not exist. Naturally she was aware of its absence, but still had sensations of its presence. Frank's view would be that it made little experimental difference whether one had the leg or not. My wife's aunt knew better because even with the sensation of the leg she couldn't walk with it since it was not present any longer. Furthermore, there are many situations where no experiment can be performed where it makes a great deal of difference whether the object be represented by sensations and be present or be not present, as independently existing.

Bernard Bykchovsky notes in his "The Erosion of Idealism": there are always such critics like the Franks who

> ..disguise their subjective idealism by linguistic trickery. They picture the fundamental question of philosophy as a pseudo-problem precisely for the reason because [sic] this removes from philosophy's field of vision objective reality.... This "pseudo-problem actually proves to be a subjective idealist postulate which serves as a foundation for the house of cards of "logical-healers", as they like to call themselves, and which collapses together with it.[9]

Frank makes no reference to Lenin's classical refutation of Mach, but states that:

> Logical empiricism..stresses the unitary character of science.. Our modern logic of science depicts the factual process of *successful knowledge* [my italics—DR] and scientific representations.[10]

Indeed? The *logic* of science depicts precisely *nothing* factual. It is a formal discipline. As Hegel says in his first sentence of the Lesser Logic in his *Encyclopedia*: "Logic as logic of philosophy develops the Idea of the whole of reality in the *abstract medium of thought* [my italics—DR]." Modern logic, in addition, since it leaves out of consideration the entire material world, fails to depict "successful" knowledge [what is unsuccessful knowledge?], and it fails to depict scientific representations or scientific presentations. According to Caudwell in *Further Studies in a Dying Culture*, "Formal logic is not [even] a law of thought, it is a rule of symbolism." But it seems beyond Frank's cunning to see that behind all this is idealistic metaphysics. Yet he does see that there is some connection between idealism and, as he scornfully notes: "..the 'intuition' of the totalitarian leader."[11] If you can substitute sensations for reality, what is to prevent you from substituting intuitions for scientific knowledge? Thus did this ideologist fight against totalitarianism that threatened to wipe out him and his people.

Such sensationistic subjectivists are by no means alone within the teeming house of subjective idealism. There are the phenomenologists like Husserl, of whom it was said, he was the ablest mathematician of his circle.[12] Marvin Farber, Husserl's student and leader of the critical naturalist school of phenomenology which attacks Husserl's transcendental phenomenology and subjectivism, has pointed out that "It appears that Husserl removes his view from any possible refutation as also do the constructionists, intuitionists and deontologizing formalists by what purport to be facts."[13] These "facts" like the so-called facts of all idealists, are not the facts of the real world, but imagined, constructed, hypothesized, hypostatized and intuited. This does not make them useless or uninteresting—only not facts in the usual sense adopted in the scientific enterprise with the possible exception of some intuited facts. Farber makes the following judgment on the idealistic tendency that separates the intellectual activity from the material considerations:

> A philosopher....may believe that he is grand and noble in his flight from reality, in his absorption in an honorific realm of pure idea, removed "in principle" from the "lowly'' world about us. But that action may be construed as a renunciation of the obligation one has towards society as well as himself.[14]

We may turn to a brief examination of the subjectivist economists to gain new insight into the large arena of subjectivism. Three years after Karl Popper's decision to undermine Marxism (1919), Eugen von Bhm-Bawerk's *Capital and Interest* (1922) appeared (eight years after B-B's death). But even earlier, from 1886 onward, it was B-B's dearest wish to destroy the influence of Marx on Continental economics. Marx was doubly dangerous because he continued classical (essentially objective) economics and foresaw the collapse of capitalism. (100 years later, bourgeois economists in the United States are predicting the collapse of capitalism in Western Europe. Edwin McDowell in the New York Times of 4 March 1984 discusses "deindustrialization" and *Nationalized Companies : A Threat to American Business* (1984) by R. Joseph Monsen and Kenneth D. Walters).

As with Popper, there is in B-B an overriding, especially in Austria a country "of greatly retarded capitalist development" in the words of the economic historian, Eric Roll. Austria produced two schools of economics (and value) with close ties to each other. Both of them were amiably supported by important British economists. Together the Austrians and the British manufactured subjectivist economics. Among these creative artists were S. Bailey, Stanley Jevons, N.W. Senior and Alfred Marshall in Britain and Menger, Wieser and B-B in Austria. They influenced many economists of lesser note yet of fertile imagination.

Classical economics had been primarily objective, but beginning in the 1820's subjectivism gradually infiltrated classical economics, reaching a peak around the 1870's. It is no accident that subjectivism rose at just the time Marx picked as the beginning of major systematic crises in capitalism (1825). To attempt to meet these crises without modifying capitalism in any structural way, what N. Bukharin called, "the rentier economics" appeared and has not reached its end even today. What was paramount in its outlook was its capitalist apologetics and declining interest in production. Petty, Smith and Ricardo put major emphasis upon production of exchange goods while the Physiocrats put emphasis upon the production of agricultural goods.

On the view of Robert B. Reich of Harvard University, the interest in production has recently continued to decrease as he noted in a letter to the editor. It was entitled "Expropriations", published in *The New York Times* (29 November 1982). This is not to say that there was any new concern for production in the years to follow. Instead of looking at the world capitalist system, Reich confines his attention in this letter to the United States. He says:

> When productivity began slowing a decade ago, many Americans discovered that the simplest way to preserve their living standard [during rising inflation] was to expropriate a portion of someone else's wealth. But this "begger thy neighbor" tactic is destroying our capacity to work together productively and thereby intensifying the decline.

Although this difficulty arises quite logically from the nature of capitalism, Marx did not predict this particular consequence of the internal contradiction of world capitalism. He did predict greater miseration for the working class which has been mentioned in other sections of this book *vis a vis* an unparalleled unemployment.* The breakdown and collapse of many industrial cities (I have lived in and observed one for the past twenty years), the geometric increase of both blue and white collar crime, the increased militarization of the system and so forth are other signs. Marx could not foresee that the results of the development of the irrational colossus would be worse than he had imagined—especially in the depredation of the general environment. The first and second World Wars of capitalism were of course unpredictable in their ferocity and waste.

The rentier class which controls the economic decisions of capitalism as a whole, in so far as it is controlled or controllable, was much less interested in production than in "the disposal of the income from its investments" in the words of Eric Roll in his *A History of Economic Thought* (London, 1973). Veblen called this rentier class "absentee owners" in his now classic *Absentee Ownership and Business Enterprise in Recent Times* (New York, 1923). Veblen, however, had a limited outlook on how the industrial system had changed and missed the warning signs exhibited in Britain. These signs now hover over the U.S.A., West Germany, Japan. A major warning sign is the lack of concern for production by former leaders of aggressive production capitalism. Now the aggressiveness is focused on take-overs, foreign investments and so forth.

With the loss of attention to production in Europe the subjective real-cost theory followed. Labor power as the criterion of value was translated into "subjective sacrifice". The structure of society as basis of theoretical analysis changed to an agglomeration of individuals analagous to the activity of simplistic atomic physics. Stanley Jevons in 1871, one of the second generation of subjective theory, cast in mathematical terms, the basic principles of subjective marginality.

First of all, this marginality was dependent on "the great springs of human action—the feelings of pleasure and pain." Value was expressed in terms of utility. This revival of Hobbesianism with an accretion of Bentham's pleasure-machine struck Jevons as being revolutionary. It occu red to few that what utility most probably measured (if anything) was price rather than value—what one would be willing to pay to get something. Weighing of motives in subjective theory made analysis difficult if not impossible. Nevertheless, like an English bulldog, Jevons stuck to it.

Carl Menger (1840-1921) took over the utility and subjectivist problem from Mill. He began by announcing that economics must stand on the firm foundation of individualism, flying in the face of several million years of human social experience. Upon this rock Menger tried to build a subjective theory of value absent of hedonism. This led to his attempt to describe volition without referring to pleasure.

Leon Walras (1834-1910) took over the problem from Menger. According to Roll, Walras was "led to his economic analysis by a desire to build up a strong case in favour of *laisser faire*, in answer to an attack by a follower of Saint-Simon." Saint-Simon, we remember, had the cheek to favor socialism, which was anathema to all the economic subjectivists. Yet, we also remember Walras less for this than abandoning the search for the origin of value in favor of constructing a purely formal theory of functional interdependence. If you are nervous about giving priority to any given factor, you come out for interdependence.

After the early Austrian school vanished leaving a residue of Menger utility plus subjectivism, the new Austrian school appeared. It was dominated by two of Menger's disciples: Friedrich von Wieser (1851-1926) and Eugen von Bohm-Bawerk (1851-1914). Although these two subjectivists added little to what Menger had already discovered, they attained greater publicity. They did go beyond Menger in providing purely formal subjective evaluations. Wieser provided the bourgeois world with the term *Grentznutzen* (marginal utility) in 1884, the year that the U.S.A. established the U.S. Naval War College at Newport, Rhode Island. For Wieser utility is the cause of value. All economic choice involves some sacrifice so that one must choose between income and leisure. In what sense this is not rentier economics remains for Eric Roll to explain since it was he who clucked against the notion that subjectivist economics was rentier economics. It has obviously little to do with the working man who represents more than ninety percent of the population.

Marginal productivity analysis of wages was not long in coming. Alfred Marshall, according to Schumpeter, succeeded here with F. Y. Edgeworth quantifying the results which were based on subjective (although allegedly objective) factors. Americans were adopting it by 1888. By 1934 econometrics burst forth on the topic of productivity analysis of wages (in the midst of the worst capitalist crisis up to that time). This removed from the market-place to advanced mathematical theory the central concern of working people.

A major presupposition in the ongoing work of marginal utility was that although "psychic reality" (Schumpeter's phrase) could not be measured directly, it was alleged to be measurable indirectly (Marshall)—by effects (the operational principle). This indirect method depended on "ordinal utility". This phrase referred to the ability to arrange subjective preferences without assigning to them magnitudes. It amounted to my pointing to a group of fruits in order of their tastiness to me. Persian melons, first; honeydew melons, second; papaya, third and so forth. This subjective construct Pareto tried to give a quasi-objective status. One gets a feeling that he is not far from what in the twelfth century would have been hierarchical theology.

HOW TO TELL SCIENCE FROM NON-SCIENCE ACCORDING TO THE IDEALISTIC PHILOSOPHERS OF SCIENCE

An important step for subjectivist idealistic philosophy of science was to decide how their science can be differentiated from the other's non-science. The logical positivists had stressed the difference between science and pseuodo-science (what they called "metaphysics"). But this attempt failed as Weinberg[15] and Cornforth[16] pointed out. Thomas Kuhn separated science from non-science in paradigm development such that one paradigm either modified or replaced another one depending upon whether the change was modification or revolution. Counter-revolution generated by ongoing research provided the knowledge for either modification or for revolution. But the agent of the change, the so-called scientific community, according to Cecil Gutzmore,[17] is so vaguely described that it does not allow separation of the scientific community (brotherhood) from religious communities or even criminal fraternities. Indeed, says Gutzmore, Kuhn's criertion"..also fits with remarkable accuracy the description offered by certain anthropologists of the operation of witchcraft and magic.." Gutzmore, citing Evans-Pritchard's *Witchcraft Oracles and Magic Among the Azunde of the Sudan* (1936), notes that "'Faith and scepticism are alike traditionally" touchstones'[8] as to whether the community is scientific or not. Kuhn's view, largely antedated by R.G. Collingwood's "absolute presuppositions" (1939) and Bachelard's "epistemological breaks", simply fails since it cannot distinguish between science and religion. I don't know how far this can be traced in modern times, but Hegel in the "Lesser Logic" of the *Encyclopedia* makes reference to it. If Kuhn had further followed Collingwood's (and Hegel's) lead, he would also have been quite agreeable to seeing no sharp division among science, religion or philosophy except perhaps in their treatment of the "infinite."[19]

Before Kuhn, Karl Popper had filled the breach with his doctrine of falsification, and Imré Lakatos with his theory of scientific research programmes. Falsification was introduced into economic discussions by T.W. Hutchison in his *The Significance and Basic Postulates of Economic Theory* (1938).

THE CUL-DE-SAC OF FALSIFICATION

Thomas Kuhn, who admits in *The Essential Tension* (Chicago, 1977) not only his general agreement with, but his admiration for, the work of Popper, disagrees with Popper's view of falsification. Popper's view is that "falsification" or "refutation" are "what happens when a theory fails in an attempted application" according to Kuhn. Hence, Popper opts for logical falsification which Kuhn cannot accept.

Hilary Putnam in his paper, "The 'Corroboration' of Theories," says that "The Law of Universal Gravitation is *not* strongly falsifiable at all; yet it is surely a paradigm of scientific theory." As a matter of fact, Putnam says, "Popper's doctrine gives a correct account of neither the nature of scientific theory nor of the practice of the scientific community in this case." (See *Mathematics, Matter and Method*, Philosophical Papers I, 2d. ed. (Cambridge, Mass., 1979), p. 257.

A.P. Polikarov considers Lakatos' "programme" to "give a dialectical direction to certain ideas of Popper," including a more palatable form for falsificationism. Kuhn admires Lakato's permissible rejoinders to mathematical refutations. Logical falsification will play a small role when it comes to determining whether one has got the truth of an objective situation.

Yet, Polikarov in *Methodological Problems of Science* (1983) states that the continuity between classical physics (mechanics and electrodynamics) and relativity physics makes Popper's falsification (especially the naive brand) "unreliable."

Gutzmore's attack on the two views of Popper and Lakatos begins by his setting down the following scheme:

(i) Dogmatic or Naturalistic Falsification (DNF) = $Popper_0$
(ii) Naive Methodological Falsification (NMF) = $Popper_1$
(iii) Sophisticated Methodological Falsification (SMF) = $Popper_2$ which leads to the theory of Scientific Research Programmes or $Lakatos_1$

Lakatos argued that the real Popper, or NMF=SMF unlike the untenable DNF Popper, was rejected by the real Popper himself (which Kuhn noted in 1977). But we shall soon discover, with the help of Gutzmore, that none of the Popper positions is tenable. Gutzmore claims that Popper continued to believe in DNF until 1959 or perhaps even later. Gutzmore then asks why Popper ever raised the issue of falsification in the first place and answers the question as follows.

> Falsification was developed in 1919 by Popper because [he says] "..it solved for me an intellectual problem which had worried me deeply, and..because it had obvious practical (for example, *political*) [my italics—DR] consequences. (1963, p. 39, ref. to Popper's *Conjectures and Refutations*). In this respect it fits closely Althusser's criteria for the mode of production of ideology.

Gutzmore refers to the ideological function of Popper's position because, he says,

> The doctrine has given good political/ideological service to liberal capitalism [not to mention illiberal capitalism—DR] that in the Cold War World it provided apparently adequate justification for the sort of political repression which masquerades under such labels as "piecemeal engineering." This means that an open abandonment of Popper is not on: Hence Lakatos' rescue operation.[21]

To step back for a moment, we find that the essence of Popper's first doctrine, namely DNF/Popper $_0$ is that only those theories which are falsifiable are scientific and that falsification is the goal of scientific activity. The guarantee of falsification is a belief in the infallible empirical level of reality. This is also called "factual" and "basic". What the objective reference is of this word we are not sure. Perhaps it is not what the American logician W.T. Parry refers to when the states that "all true propositions denote the same thing, viz. the actual world; false propositions denote nothing." ("In memoriam : Clarence Irving Lewis (1883-1964), *Notre Dame Journal of Formal Logic*, XI, 2, p. 135).

Nor need he be referring to the facts of nature which are always consistent (even though the "facts" of logic need not be as was noted by Thomas J. Hudson in *The Law of Mental Medicine* (Chicago, 1903).

According to Popper, the observing experimenter works at the factual and basic level (in Popper's sense of factual and basic) whereas the theoretician makes conjectures" and "guesses." A theoretician must *specify in advance* any experiment such that an adverse result would falsify the conjecture. *In the name of Nature*, the experimenter disposes of conjectures with the aid of basic empirical facts together with a well-corroborated falsifying hypothesis. This is to say, an explantation of why the fact falsifies the conjecture. The reader should be reminded that Popper assumes a dualism that the experimenter and the theoretician are not the same person. Consequently, the experimenter is working the basic empirical facts while the theoretician is making conjectures. There is no appeal in this doctrine against experimental falsification. Difficulties of the doctrine include the following :

(1) No such line as is specified can be drawn between experimenter and theoretician.
(2) It is self-contradictory in that while being based on the supposed impossibility of proof its conception of disproof turns out to be itself a doctrine of proof (in which infallible Nature can shout an emphatic NO to any conjecture).
(3) It is absurd to expect genuinely scientific theories to specify in advance their condition of failure. If they could, then they would be immediately changed.
(4) It is ridiculous that genuinely scientific theories must forbid certain observable states of affairs. According to Lakatos it is the most admired theories which fail to forbid certain observable states of affairs.[22]

It might also be mentioned that the bases of (3) and (4) provide for Lysenkoism in science.

We should note in passing, that regardless of the outcome of Popper's falsification, he contributed to the recognition of the impact of metaphysics (and ideology) on philosophy of science as did Einstein when he said that physics is mostly metaphysics. Popper's contribution here is neutralized to some extent by Popper's unstated principle of the rejection of revolutionary theory congruent with his early rejection of histodiamat. (On this see also Polikarov, *Methodological Problems of Science*, p. 74].

Now, since this Popper doctrine is untenable, let us move on to the MF/Popper$_2$ which is a conventionalist position according to Gutzmore. It insists upon the theoretical (conventional) basis of all scientific knowledge, but at this level it is *petitio principii* (circular) since it assumes what Popper is trying to establish. It is conventional in the sense that it is social as a result of intersubjective argument (agreement) among the interested scientific community. Popper, in his search for falsifiability, talks about "decisions". In order to separate rejection from disproof, he postulates the following decisions:

(1) A decision as to which facts represent the basic or empirical statements of a theory intersubjectively agreed upon.

(2) A decision as to which of the basic statements of a theory based on facts are its potential falsifiers, i.e., which agreed upon statements are a basis for declaring a theory to be falsified. The experimenter is the chief provider of such statements.

But, Gutzmore points out, such decisions cannot be used to disprove conjectures. They can only show the conjecture to be inconsistent with a potential falsifier. Any such conjecture which is shown to be inconsistent is rejected or falsified. Falsifiability is not guaranteed by such decisions, nor is scientificity. An example of this is the set of various theories of probability. To cheer on the reader to this increasingly complicated Popper's broth, Gutzmore says that Popper's nearly unbelievable picayunishness may "do violence to the staying-power of normally patient men........" [23] trying to do normal philosophy of science.

As we might have predicted on the basis of Popper's known picayunishness, he has further decisions to examine :

(3) A decision to specify "rejection rules" for rendering statistically based evidence "inconsistent" with any particular instance of probabilistic interpretation. Theories of probability are thus rendered "falsifiable" in principle.

(4) A closely related "set" of decisions designed to solve the problem flowing from the conception of the nature of scientific theory held by MF/Popper$_2$. MF/Popper$_2$ views theory as comprised of four elements, namely: (a) universal statements or hypotheses of the character of natural law, (b) singular statements or "initial conditions", (c) singular predic-

tions, and (d) a *ceteris paribus* [other things being equal] clause. Such a clause greatly admired by formalists because it implies that no concrete knowledge must be dealt with when considering the postulate of perfect competition in economics[24].

With regard to the fourth component (d) which is always present in any research model it is virtually unspecifiable with respect to a major theory—so that the *forbidden* occurs precisely in the most respected theories. Which is to say that the *expected* fails to occur. In the case of the Einsteinian replacement of Newton, notions as apparently unproblematical as succession and simultaneity are "relativized" to complete the destruction of a part of Newtonian science. The core of Einstein's view is precisely what $Popper_2$ forbids.[25]

(5) Finally, MF/$Popper_2$ is complete when we add two further decisions which are: (e) it will be decided that a theory not rendered falsifiable by a previous decision is unscientific if it clashes with any other theory which is both falsifiable (on the above grounds) and also, at the same time "well-corroborated", and (f) a decision that theories which are syntactically metaphysics," i.e., have a logical form such as "all metals are solvent" which precludes their falsification by any single ("spatio-temporally singular") basic statement, are rejected as unscientific.[26]

Asarya Polikarov contributes the notion that formalism and experiments attempting falisification are of little use if we don't have any bright ideas (See the *Born-Einstein Letters* (London, 1971), p. 151. See also Polikarov, *Methdological Problems of Science* (Sofia, 1983), p. 56. T. I. Oizerman warned in 1982 that "Popper's falsification theory inevitably leads to a denial of the difference between a hypothesis and a scientific theory obtained by experimental confirmations and tested in practice by, for example, applying it in technology," ["Some Problems of the Scientific-Philosophical Theory of Truth," *Soviet Studies in Philosophy*, XXI, No. (Spring 1983), originally from *Voprosy filosofii*, 1982, no. 7].

To return to our Byzantine studies, Gutzmore further objects to $Popper_2$ that $Popper_2$ (1) leaves unstated the central declaration upon which this view rests, which is to say, that in 1919 Popper decided to regard science as distinguishable from non-science by virtue of the supposed falsifiability of science as distinguished from the non-falsifiability of non-science. But this violates the central canon of Popper and his disciples. Both $Popper_1$ and $Popper_2$ (the reformed and "improved Popper), by being totally committed to this principle, are left in the position of resting on the shifting sand (although sometimes fruitful) of self-contradiction. (2) Popper's project reveals a double ignorance. The first ignorance is that the 1919 Popper's decision was not based on knowledge of actual difference between natural and social science to which it had reference. Social science, which Popper declared unfalsifiable (and hence not science) was composed of science(s) about which he knew little or nothing. According to Gutzmore these included psycho-analysis, individual psychology, sociology, economics—a charge which may be easily checked by examining his *Open Society*.[27] I believe that Gutzmore's list of Popper's incompetencies should include other social sciences such as history of philosophy, ethics, anthropology and so forth. In addition, Popper is unable to present accurately the views of Plato, Hegel or Marx (probably because of his lack of training and limited perspective in "intellectual" history in the more humanistic respect). Gutzmore does not go into the relationship between "accuracy" and ideology, but such a discussion could be expanded into a monograph. Popper's accounts of the views of the philosophers mentioned above amounts oftentimes to non-knowledge. It would be a toss-up as to which accounts would prove the least accurate : Popper's or Bertrand Russell's in the latter's *History of Western Philosophy*. Russell's advantage is in his being at the very least risible and at the most original and insightful even when mistaken. It is only too easy to list the areas of knowledge with which Popper deals with considerable deficiency.[28]

What follows from these foregoing remarks is that as late as the 1950's Popper did not "have a basis of knowledge for the attempt to make unfalsifiability the key *presence* in the social sciences just mentioned—falsifiability postulated by Popper as the key presence in the natural sciences, especially physics. Both claims of Popper were apriori and prescriptive.[29]

What Popper did was to apply a model *derived from traditional logic* to certain sciences and say that it was not applicable to certain others. The basis of this error was ontological since Popper did not think in terms of an independent material world. When you move back and forth promiscuously between the formal mode and the real material mode (not Russell's specious material implication which is ideal implication) you begin to "psych yourself" into erasing the line between the ideal and material. Popper's self-revealed confusion is projected to everyone when Popper states of some social theories that they are both falsifiable and non-falsifiable. Popper ought to have noticed, says Gutzmore, that when falsifiability is defined as limited testability, both physical and social theory are equally falsifiable and unfalsifiable. Gutzmore adds: "He would have noticed too, that the attempt to generalize testability *from a logical model to science* has no basis in [acceptable] theory, that in other words, science is not traditional [merely symbolic, modal, relational] logic."[30] Kuhn recognizes this difficulty when he states that "Debates over theory-choice cannot be cast in a form that fully resembles logical or mathematical proof." (*Structure*, 2d. ed. enlarged, p. 199). The desire to replace all other models by logical models is epidemic among subjectivists such as W.C. Humphreys, R. Carnap and "pure" economists. Such approaches obviate the need for concrete application which might illuminate the spectator and make him unnecessarily critical of the formalistic activities at hand.

The third objection by Gutzmore is that the fifth decision of Popper, if it is regarded to apply to theories as a whole, entails the risk of rejecting all new theories. On the other hand, if Popper's view be interpreted as reasoning from a partial perspective, then the decision is about propositions. And, by the way, *what are* "syntactically metaphysical" propositions. How can metaphysical be demarked by grammar and word order ? How can metaphysical be differentiated from scientific by word order and grammar? Gutzmore observes that :

> Propositions either have or lack a scientific/theoretical context, to which their logical form is irrelevant. Thus the proposition "all metals have a solvent" is indeed syntactically metaphysical (no single observation will disprove it)."[31]

When we return to the "improvement of Lakatos over Popper," says Gutzmore, we discover that Lakatos rejects the notion of falsificationism which he replaces by supersessionism. On this view, a theory T is superseded if and only if an alternative T_1 has been proposed and which has the following characteristics : (a) excess empirical content over T, thus enabling T_1

to predict novelties forbidden or improbable under plain T. This should be checkable through apriori, logical, analysis and even to some extent empirically over an indefinite period of time. (b) T_1 must explain the previous success of T, so that T is included in T_1 insofar as T is not supersedable. Scientificity for Lakatos in its highest form of progressiveness is only derivable from a determinate series of theories and not from a single one. The units of such series are : (q) hard core or a relevant theory, (r) its effective rivals, and (s) its auxiliary hypotheses (a protective girdle around a hard core); and (t) its initial conditions which serve as the basis for models for research on specific aspects of theory (Collingwood's absolute pre-suppositions). The units of the series include the (x) hard core [central structure] or (y) relevant theory plus its effective rivals, (x) plus its protective belt (auxiliary hypotheses), and its initial conditions. So that x+y+z make up a research programme.

Lakatos has two levels of theory formation: first, an explanatory, theoretical level, and second, an interpretative theoretical level, both of which are called "high level". It is their function to solve the Popper-Lakatos "untenable distinction between 'observational' and 'theoretical' statements."[32] The interpretative (second) level provides the facts which are then explained on the explanatory level (first). When they clash there is an inconsistency. To accommodate himself to such inconsistency, Lakatos offers the strategic "problemshift" which has a dual nature consisting of a theoretically progressive element and an empirically progressive element. The theoretically progressive element explains and predicts some previously unexplained or unexpected or forbidden facts (!), whereas the empirically progressive elements, present—the excessive empirical content—are confirmed on the interpretative level.[33] In my opinion "inconsistency, "unexpected" and "forbidden" are taken too rigorously. With a more dialectical "letting it hang-out" like a dress shirt not yet tucked into the trousers these at-first unexplained elements should be treated with more leniency and openness following the principles of histodiamat before they are rigidly interpreted by scientists clinging to positivistic and metaphysical skirts.

Lakatos says that "......the theoretically progressive element is a consistent demand of genuinely scientific practice while the empirically progressive element is (almost by defintion) only an intermittent requirement."[34] It is a stipulation for continuous theoretical growth in science. We must reflect from time to time that the emphasis on rigor may well impede an emphasis on growth. There can be *rigor mathematicus* or *rigor scientificus* as well as *rigor mortis*. Frege once said that "It would perhaps not be wrong to say that the laws of logic are nothing but the development of the word 'true'." [See *Nachgelassene Schriften*, ed. H. Hermes et al, (Hamburg, 1969), p. 3]. The theoretical progressive element is to be substituted for unity and beauty although not simplicity, since simplicity is often a spurious consideration when we realize that systems of explanation involved different kinds and levels of difficulty. For Lakatos, a problemshift may be either progressive, that is scientific, or degenerating. To be progressive is primary because it guarantees"'......the historical fact of the relative autonomy of theoretical science.'"[35]

Gutzmore's criticism of Lakatos is that Lakatos' study (theory of scientific research programme) is simply Popperian without falsification. It is similar to Kuhn's theory also. Noting that Popper grants that the pre-Marxian social science is now inconceivable (*Open Society*, II, 82), Gutzmore asks what contemporary social theory has greater explanatory theoretical power than historical materialism ? We remember that it was Popper's initial thrust to show that Marxism is unscientific because it cannot be falsified. With falsification shown to be a useless stricture, however, Popper's argument vanishes. Then the question rises, what other arguments could Popper and his followers raise to show that historical materialism is unscientific ? They produce nothing, Gutzmore maintains, even though Lakatos was expected to step into the breach.

The trouble with Lakatos is that his view SRP (Sci Res. Progr.) is a totally descriptive theory consisting of a dialogue between two modes of description. One mode is a rational reconstruction while the other mode is actual history. The dialogue takes the form of a criticism of the one for a *lack of historicity* and the other for a *lack of rationality*. It will be apparent that such a procedure can amount to little else than a more or less interesting juxtaposition of schematic and detailed descriptions.[36]

Even Lakatos is dubious about its justification, for, as he said in 1962 :

> My respectable historian colleagues sometimes say that the sort of "rational reconstruction" here attempted is a caricature of real history; but one might equally well say that real history is a caricature of its rational reconstruction.[37]

Lakatos seems unaware that this lets the cat out of the subjective and formalistic bag. He implies in the above quotation that it is just as important or proper to have coherence as to have correspondence when dealing with the independently existing real world. It isn't clear whether he realizes that correspondence provides the meat for the epistemic soup. Without the meat there is no constituent. For Lakatos, it appears that history is just as much, or perhaps more, the work of historians as it is the work of society in its action on and with nature. It is of course exceedingly clear to anyone sensitized to the history of modern subjectivism and the de-ontologizing of thought since Kant, that Popper, Lakatos, and sometimes even Kuhn are of a similar subjectivist humor. I say that despite the view of Assen Kozharov that Popper is now an objective idealist. According to Kozharov, Popper has moved from positivism [subjective idealism to an "objective idealism of a Platonic type" in his *Monism and Pluralism in Ideology and Politics* (Sofia, 1974]. But we maintain that ontologically, objective idealism must reduce to subjective idealism since its "objective realm" is also subjective. It follows that Berkeley's *deus ex machina* is subjective also. The third-world of Popper is still only in consciousness.[38] Simply because philosophers write and speak as if their subjective realm (like Plato's forms or Santayana's essences) were objective is no reason to believe them or humor them. If the historian of philosophy claims that a distinction between subjective andobjective idealism is customary, then I respond by saying that custom also is not inviolate.

Lakatos admits that "rational reconstruction" takes place in the world of special ideas, which is to say, in Plato's and Popper's "third world". This world is distinct from the first (material) world and the second "consciousness" world. There is matter, mind and something more according to Popper and Lakatos.Although this is perhaps the most serious charge against Lakatos, that he falls into subjectivism and Platonism, we must turn to the generally praised views of Thomas Kuhn to complete the picture of current subjectivist philosophy of science.

MICHALOS INTERLUDE

Let us turn briefly, by way of preface, to *The Popper-Carnap Controversy* by Alex C. Michalos. We are assured in the first sentence of his book that the controversy is not about substantive issues. Michalos says :

> In the *Logic of Scientific Discovery*, Karl Popper introduced and defended the view that the acceptability of a scientific hypothesis is directly proportional to its logical probability .[39]

I doubt that probability has much to do with logic in this context. One would have hoped that he had said "in proportion to its congruence to reality". But he didn't. That leaves us in the usual formalistic situation without reference to ontology and hence to the external world. We are confronted with a *fait accompli* that there will be no discussion of the relationship of scientific hypothesis to the independently existing material world. Instead we shall be confined, as usual, to the squirrel cage of formalistic "determination". In this situation freedom is not the insight into necessity; it is the critical understanding that formal "necessity" has been substituted for real necessity in nature or society.

KUHN'S STRUCTURE OF SCIENTIFIC REVOLUTIONS

Popper believed that he could demonstrate the difference between science and non-science formalistically using a logical model. Thomas Kuhn attempts to improve on this approach as we shall now see. In his "Postscript" to the second edition of *The Structure of Scientific Revolutions*, Kuhn states that "......[t] he preceding pages present a viewpoint or theory about the nature of science...."[40] Others have claimed much more for the *Structure*, including the introduction of the word "revolution" into normal Western philosophy of science and a new emphasis on change, if not *change as development* from something to something else. A. Hall in 1954 or eight years earlier, introduced the notion of revolution in *The Scientific Revolution* 1500-1800, but it was a notion that he hedged about carefully in his concluding chapter. This chapter contained the thought that "......though it is easy to frame hypotheses [?], and consequently the whole problem of the connection between the changes in society and concurrent changes in science is involved in doubt, to all but the dogmatic Marxist," thus making it quite clear that he is incorruptible on this Cold War score. This is perhaps an example of what Freud termed *Widerstand* (resistance). When Freud was analyzing Dr. Joseph Wortis he said to Wortis : "You're always mumbling......like the Americans do. I believe it is an expression of the general American laxity in social intercourse, and it is sometimes used as *Widerstand.* "On the other hand, the American legal philosopher, Mitchell Franklin complained of (instead of *Widerstand*), *Verstellung*, that is ambiguity, ironical statement, slipperiness, when criticizing a similar stance.

Kuhn, of course, made a considerably bigger splash in North America than in Europe where he was judged within a larger context of methodologies. Yet, his work is not negligible from that larger perspective as to the nature of the scientific enterprize on the view of a number of European Marxists.

In the *Structure* we discover that for Kuhn the nature of science is such that its practitioners share a paradigm and a disciplinary matrix, forming a community. Bernal, earlier, had said that science is what scientists do, a notion that is also to be found in P.W. Bridgmen's *The Logic of Modern Physics*, where Bridgman maintains : "In general, we mean by any concept nothing more than a set of operations, the concept is synonymous with the corresponding set of operations." The community of scientists, on this view of Kuhn (and others) consists of the producers and validators of scientific knowledge. All this, however, is circular because science is always defined in terms of science and operations of science recognized through the operations of science. In the *Social Function of Science*, Bernal had called science an institution. In *Science in History* he speaks of Science as method, cumulative tradition, productive element, as well. Kuhn gives a hypothetical definition. He says "If science is the constellation of fact, theories, and methods collected in current texts [of science ?]......" once the paradigm and disciplinary matrix of science is identified (operationally), it is possible to find symbolic generalization growing out of the matrix in the form of predictions, resulting from these activities says Kuhn.[41]

All this is unsatisfactory. In such a context, the science of religion is advanced over the science of science because one can define religion as a series of rituals etc. in which man recognizes a relationship to a sacred or spiritual "being." Kuhn is in good company because Van Nostrand's *Scientific Encyclopedia* does not even attempt to define science. *Webster's Collegiate Dictionary* has two general definitions : one in which you cannot differentiate science from any other knowledge or purported knowledge and a second one which is circular. George Sarton tries to differentiate science from magic, which Kuhn does not try, by maintaining that "......science is essentially progressive....."[42] It was H.G. Wells who said : "Without the idea of progress life is a corrupting marsh." Sarton adds (I, 23), "Science is reason organized and systematically applied. Religion is a reasoned and not unreasonable abdication of reason with regard to problems which are not amenable to scientific treatment." There is no conflict between science and religion, says Sarton, except where they trespass on each other. As A.N. Whitehead said in *Adventure of Ideas* "Religion is the last refuge of human savagery." In discussing this with him in Cambridge, Massachussetts in 1945, Whitehead gave me the impression that he sometimes saw little difference between science and religion, again reminding me of Einstein's statement that physics is mostly metaphysics. Both men were sophisticated dialecticans in many areas. Harry Wells showed the weaknesses of Whitehead in *Process and Unreality* (1950), especially Whitehead's dogmatic adherence to "persistence" in metaphysics. This was probably the result of his early emphasis upon mathematics and perhaps the eternal objects of Plato. His belief that most of European philosophy was a series of footnotes on Plato left off the idea that many of the footnotes were *against* Plato. Whitehead said that "The aim of science is to seek the simplest explanation of complex facts" in his *Concept of Nature* (1920). Despite his great powers as a simplifier, Whitehead was said by Bertrand Russell to be muddleheaded. I found him a charming and witty philosopher who had the advantage of having a philosophical wife who was also a good cook and energetic political thinker. Like William James he tried to understand Hegel through Bradley and hence missed the most important philosophical development of the nineteenth century. Whitehead says that "Truth is the conformation of Appearance to Reality," in *Adventure of Ideas* (1940). This implies correspondence. The "blythe spirit" school of twentieth century philosophy in Britain in the person of F.P. Ramsey claimed that to assert that "x is true is merely to assert x." If this were only true in the real world—for it would eliminate thousands of lawyers, criminals and tax investigators. Ramsey combined the virtues and vices of the "blythe spirit" school since he had the virtue of simplification but the vice of oversimplification.

The attempt to de-emphasize or even degrade truth is a common feature of declining civilizations. Various dodges are paraded in place of truth. Gertrude Ezorsky in her "Performative Theory of Truth" (*Encycl. Philos.*, 5-6, 89) correctly sees that "True', unlike 'yes' has the force of adequate evidence." Hence "x is true" contains a judgment that disappears in just plain "x".

But before we leave the topics of science and truth momentarily, I should like to set down a definition of science that has certain virtues lacking in Western European definition frequently held. "Science is a form of social consciousness which represents a historically developed system of knowledge whose truth is verified and constantly made precise in the course of society's practical experience." This definition by M. Rosenthal and P. Yudin in 1967 seems to lack very little. It is contained in their *Dictionary of Philosophy* published in the same year.

We return now to a review of Kuhn's popular outlook. When a scientific community is involved in a "revolution" it participates in a "......reconstruction of group commitments," Kuhn states. It may be a neat summation, but again, not merely applicable to scientific activities. Such revolutions, Kuhn says, are heralded by crises, which may be generated by the work of the community of scientists, by a development of new instruments or by the assimilation of new laws. He might have added that revolution might also occur by a major change in the mode of production created by shortages, warfare, prodigies, holocausts and the like. So far, we have seen Kuhn dealing with science (defining it operationally) in the circular terms of how scientists behave. In *The Encyclopedia of Philosophy* Peter Caws creates the same difficulty using the term "science" in his definition of scientific method. He does tell us, however, that scientific method in its lowest common denominator, "......can amount to little more than fidelity to empirical evidence and simplicity of logical formulation." The "logical" part would appear to have about the least relevance when we go beyond the Popperian formalistic (and anti-social) parameter. Another joker is that we don't know what Caws means by "fidelity". "Fidelity" could easily turn out to be nothing more than coherence which will not take "science" beyond magic and religion. Even nonsensical fidelity could fit the prescription—as in fatuous games—usually made up by people who enjoy the free play of coherence without any correspondence with reality. In its own way, *Alice in Wonderland* might be seen as a coherent system with a good bit of "fidelity."

Kuhn's chronic circularity is rooted in the same difficulty facing Caws. He is also treating science formalistically and contemplatively without reference to the independent external world. Missing is a tie to the history of science and technology which can give concreteness to the discussion of what science is.

Gutzmore's assessment of Kuhn's alleged achievement is the following :

> Kuhn appears to offer little more than a partial description based on certain historical instances of change and growth in what is traditionally called "science".. shows how certain theories developed, came to dominate fields and how they were themselves replaced.[43]

But of course Kuhn has missed much of the work done in the productive historical process—at least he neither describes nor mentions it. The kind of work he set for himself, it seems highly likely, could perhaps be adequately done by an historical materialist. What is of interest and of note is that Kuhn's description (he has no definition) of science or scientific activity also fits the activity of such other organized groups as religious communities and criminal fraternities. It even fits remarkably well the description of the practice of witchcraft in tribal societies where faith and skepticism are both traditional. It might have helped Kuhn if he had paid attention to Sarton's mentioning that science was somehow connected in some way with progress. That might have entailed, however, some kind of philosophy of history which presumably is not on. But he would not have received much help in the following definition to be found in Sarton's *History of Science.* For in this work Sarton says :

> What is science ? May we not say that whenever the attempt to solve a problem is made methodically [methodical fertility rites?] according to a predetermined order or plan [rainmaking dance?], we are witnessing a scientific procedure, we are witnessing the very growth of science?[44]

Well, within the broadest context of social development, perhaps ? But the context is necessary for understanding the role of magic and how magic becomes science. Sarton properly raised the question as to what is science. Yet, as he handles it, I should like to respond to his question as my old friend Professor C.H. Chen, the Aristotelian scholar, often did. He said to such questions : "Yes, I don't think so."

Kuhn does not separate science from primitive ritual or contemporary ritual. Kuhn himself makes reference to religious dogma [ritual]. He refers to theological "progress" in one place as if he can only conceive of moving towards a goal in a theological rather than scientific framework—if he is able to differentiate them.

Obviously one cannot treat science quite as a-historically as Kuhn and the logical positivists, logical empiricists, mindless empiricists and subjectivists would like. They will say that Kuhn does indeed treat it historically simply because he mentions a number of earlier discoveries of science and that the title of his earlier work, *The Copernican Revolution* jolly well proves it. In this work he traces the history of planetary astronomy. A history is surely a history. Yes, but what kind of history is it ? Is it a history of a form of social consciousness which represents a historically developed system of knowledge whose truth is verified and constantly made more precise in the course of society's practical experience ? Is it coming from somewhere and going somewhere ? Is it not only dealing with universality but also necessity and objective truth ? Religion and magic give distorted pictures of reality whereas science gives increasingly true pictures of objective reality. You have to get critically beyond science to know that. And we have to know that in order to adequately describe and define science.

The next item of importance, after the discovery that Kuhn

does not define science, is his view of epistemology and ontology. That he recognizes ontology as a fact of historical discussion appears at first glance to be reassuring. But his subjectivism, his neo-Kantianism, is at once apparent in the following passage:

> There is, I think, no theory-independent way to reconstruct phrases like "really there"; the notion of a match between ontology of a theory and its "real" counterpart in nature seems to me illusive in principle. Besides, as a [sic] historian, I am impressed with the implausibility of the view....[45]

This puts me in mind of Henri Poincare's (1854-1912) dicta in *The Value of Science* that "......not only science cannot teach us the nature of things; but nothing is capable of teaching it to us... ." But Kuhn is academically quite correct in saying that there is no theory-independent way to reconstruct phrases like "really there." Immediate sense information is generally prior to theory [do primitive people have theory ?] although it may imply an implicit theory to cogitate about it in a sophisticated way. Yes, what we can say about "really there" depends upon our epistemological theory which is enmeshed with our ontology and metaphysics (dialectics). The original data of theory, however, come from the various relationships among sense experiences, reflection, opposition, ideology and other theory and so forth.

Kuhn's next statement is theory-dependent. Kuhn is impressed by the implausibility of the view that there can be a match between ontology and its real counterpart in nature—the material world. This is his Kantianism. The independent external world cannot really be known. Why not ? Because Kuhn accepts the epistemology of the Kantians that we cannot really know the external world because we are inhibited by (a) the structure of the human mind (how does the animal mind know the external world ?) as allegedly discovered by Kant in the eighteenth century. And (b) the unco-operative nature of the refractory material world. The allegation that a match between ontology of a theory and its "real" counterpart in nature is "illusive in principle" is merely Kuhn's way of saying that he accepts the Kantian view of the unknowability of nature. So it is not illusive in principle at all. It is merely *illusive for Kuhn* because he accepts Kantian agnosticism based on Kant's idealistic dualistic anti-scientific epistemology. He erroneously expects the rest of us to agree with him that nature and the human mind are what he says they are. When in *The Essential Tension* Kuhn says that "theory should be accurate" he does not say accurate with relation to what, except "experiments and observations" as if these were independent of philosophical positions (Kantianism). When Kuhn discusses "objectivity" versus "subjectivity" he thinks of subjectivity in a very narrow sense as if it had mostly to do with idiosyncrasies in taste like "I like peanut butter sandwiches", or "I like wherewolf movies". That it has anything to do with materialism versus idealism is quite beyond his understanding. Ultimately objectivity for him will be related to "accuracy" and "consistency." Accuracy depends upon one's view of what nature is and how one knows the external world; consistency is a formalistic logical notion that may hold in games like bridge or canasta so is certainly not a seminal or basic constituent as to the objectivity of a theory.

It is difficult to discuss all this without also going into the aetiology of scholarship like Kuhn's which has as a component the criterion of simplicity, yet on the one hand it leads to elegance and on the other hand to gross oversimplification in understanding the history of science, mathematics, logic and philosophy. Both Leibniz and Bertrand Russell are outstanding examples of what happens to the philosophical thinking of intellectuals who are frequently absorbed in mathematics and logic, although Russell, among his other contributions, believed that Kant's apriori philosophy was destroyed by the discovery that a defect in Euclid made Kant's epistemology unnecessary as well as false. In his last years Russell came to the realization that the external world exists independently of thought although as a young man he began by accepting Plato's idealism and then had various assignations with Wittgenstein, logical empiricism and pragmatism all of which encouraged his "natural" subjectivism. Fascinating as is the century-long philosophical pilgrimage of Russell, we turn now to a more minute study of Kuhn's agnosticism.

KUHN'S AGNOSTICISM

Supposing that we are never sure if we know nature, what

is it that we know when we construct theories about nature ? This is a fair question to ask any agnostic. We put the question here to Thomas Kuhn. Furthermore, in addition to this first question, why is it that "well-constructed theories" enable us to describe, explain, control, co-operate with, nature ?

Kuhn's answers to these questions are not given in his books. But we can have some idea of them. First, Kuhn may say that we don't really know nature at all, or that we may know nature, but we can't be sure. It appears that he believes that the prudent thing to do is to act like he knows nature, even if on his view he can't be sure. At the same time he should make it clear that we don't know really whether scientific knowledge is getting us closer and closer to knowing nature. Yet, if we continue this line of reasoning it seems that we go back to Hans Vaihinger, another neo-Kantian, to his *The Philosophy of As-If*. Vaihinger came by his as-if-ism honestly. His father was a theologian and Sanskrit scholar who made the Mahabharata, a great Indian epic, a special absorption. In his university studies, Vaihinger during his first year studied ancient philosophy, in the second philosophy up to Kant, the next from Kant to Hegel and finally beyond Schleiermacher. Vaihinger claims that Kant freed him from "the pressure of the material world." Vaihinger also got hold of Eduard von Hartmann's *Philosophy of the Unconscious*. Combining the agnosticism of Kant with the unconsciousness of Hartmann presumably led to a philosophy of as-if. Then too, there was a strong imprint of Schopenhauer who believed that *the world was his idea*, common enough among the nineteenth century idealistic Germans.

This fictionalism seems to have affected Kuhn more than he ever realized. An interest in Kant affected Kuhn (reinforced by the historical fact that it was largely Germans doing the physical theory of the nineteenth century, reinforced at the rise of quantum mechanics as interpreted by Schrodinger. Schrodinger, when he discusses wave mechanics puts emphasis upon the role of the observer as did Hertz and Heisenberg. Theories of relativity also made much of the observer in determining the special field under consideration. So subjectivism took a quantum leap at the time of the special theory of relativity in the second decade of the twentieth century. Einstein, although flirting with idealism and subjectivism, ultimately came out strongly for the objective and materialist interpretation of physics and nature.

Another factor making for subjectivism leading to Kuhn's own milieu was the inability of the Newtonian mechanics to accommodate to the Lorentz transformation. The Marxwell equations on the other hand, are covarient under the Lorentz transformation (and with the constancy of light) which implies that the equations of electrodynamics have the same form in all systems in which the equations of mechanics hold. What is important for us here, however, is the hypothesis upon which the foregoing is based, namely, stating that the velocity of light is the same as measured by any one set of observers moving with constant relative velocity.

The role of observers brings in the probability of subjective judgment. It is at this time also that questions are raised concerning the personal equation in psychology and the other social sciences.

The general theory of relativity attempts to relate the measurements (potentially subjective) of observers who are accelerated relative to each other and hence not in an inertial system.

Transforming these considerations to the social sciences by analogy, we begin to notice the effect of the idea of observers as scientists observing nature, switching to considering them not only as observers but also as centers of ideology (prejudices). But where it was possible in mathematical physics to come to a central focus of constancy (in the speed of light), under social sciences the closest to this constant was the histodiamat presupposition which gives a certain stability to all observers because they are all subject to the laws of the productive system and its relations.

When Einstein entered the political economic arena it is not strange that he was attracted to socialism because of its underlying equalitarianism as well as its greater theoretical elegance which is responsible for its use by all progressive revolutionary peoples. When compared with the jumble of half-analysed capitalist beliefs which are unrelated and riddled with sloppy views of human nature, histodiamat appealed to progressive intellectuals. "Observers" in mechanics and electromagnetism became (ideologists) in the social sciences. It appeared that when histodiamat principles were accepted the social sciences progressed as was admitted by Popper himself. If histodiamat was rejected various subjective melodies (cacophanies ? variations on a Kantian theme) arose; or appeals to human nature which cultural anthropologists and archaeologists were likely to flatten. All of these various subjectivist views in social science were general referred to as "bourgeois social science" which included agnosticism—up to then considered uncritizable because of the generally simpliestic anti-theological viewspoint of many pluralists and naturalists by the twentieth century.

Under exotica appeared abstractionism, Bovaryism, existentialism, Legal Marxism, neo-Kantianism, phenomenology, phenomenalism, personalism, pragmatism, symbolism, voluntarism and certain forms of linguistic philosophy, positivism, physicalism, formalism and intuitionism, an incomplete list.

Two streams of subjectivism interlaced. The earlier going back to the rising intellectual activity of the 1870's such as Hartmann and Vaihinger, Mach on reality as made up of sensation, Freud on the unconscious, formalism in mathematics, logic and logical positivism and logical empiricism, symbolism in literary criticism tied up with sensationalism and so forth. There was subjectivism also in the physical sciences connected with observers, experimenters, scientists constrained to choose the dualism of science on one side and the rest of social life on the other. And interwoven with these two trends of subjectivism were the traditional theological considerations about knowing the really real word and knowing soul and matter or mind and matter. Politically there was a focus on class struggle because of the failures of capitalism beginning in 1825, social revolutions, progress, social evolution, imperialism, and biological considerations impacting on evolution and revolution.

Indian thinkers were caught up in all of this without knowing much about the history of its emergence. They were not always helped by Western interpreters who were ideologists to be sure, but not sharply acquainted with progressive developments and having no philosophy of history. Philosophies of historyes such as Spengler's and Toynbee's turned off the sophisticates who found that these humanists were outside the mainstream of scientific advance. That tended to leave intellectuals in the arms of Kant or Marx and Engels.

To return to fictionalism, we can say that it was motivated partly at least by pragmatic considerations. According to Vaihinger, "What we generally call truth, namely the world of ideas, which corresponds to the external world, is only an efficient error." Indeed ? Then why not an inefficient error if it contributes nothing ?

Vaihinger also says : "Our understanding of what we perceive is subjective; what is subjective is fictional; what is fictional is untrue; and what is untrue is error."[46] In his "Preface to Shakespeare," Dr. Samuel Johnson (1709-1784) concludes : "The mind, which has feasted on the luxurious wonders of fiction, has no taste of the insipidity of truth."

Our two examples of subjectivist philosophy (also called difference-philosophy or pluralism by the American legal historian, Mitchell Franklin) are a paradigm of the fix one gets into following the principles (or lack of principles) of the neo-Kantian epistemology and ontology.

R.D. Ranade has discovered that Vaihinger's outlook is similar to the view of the legendary Indian sage Yājñavalkya in the Bṛhādaraṇyakopaniṣad (4th Brāhmaṇa). Ranade asserted one important difference between the neo-Kantian Vaihinger and the Upaniṣadic sage. It is that :

> While Viahinger's fictionalism is sensationalistic, Yājnavalkya's fictionalism is Atamanic [atman=soul]. When Vaihinger was asked as to what remained when all his facts became fictions, and when the challenge was thrown at him that all fiction thus become [sic] fact optimi juris [sic] Vaihinger retorted by saying that there were two entities about the reality of which we might say were certain, namely the flux of sensations, and the laws of identity and contradiction.[47]

For Vaihinger there is double-truth. For Yājñavalkya there is the higher and the lower knowledge : *avidyā* and *vidyā*. In this sense, Kant is closer to the Hindu sage, according to Ranade, then is Vaihinger. Kuhn, with that irresponsibility that hinges on the usual absence of dialectical thought, lets others spin out the implications of his agnosticism.

It is high time that the view be reexamined and rejected that gives the impression that anything skeptical and agnostic is sophisticated and everything non-skeptical affirmative is naive. It depends. Many of my moderately prosperous students are more skeptical about the value of science and technology than they are about the value of Eastern religions. Does this make them sophisticated ? My female students are more skeptical about the independent existence of matter (especially as seen by Lucretius) than about the independent existence of spirit. This does not mark them out as particularly knowledgeable as far as I can see. That Kuhn has doubts about really knowing nature indicates to me that he has had a thorough training in agnosticism and idealism. When he was trained, neo-Kantianism was in the scientific as well as philosophical saddle. The thinkers who impressed him the most as mentioned by him in his Preface to *Structure*, include eighty-five percent idealists, at the least. Kuhn's skepticism about whether the real can ever be known is indicated in other parts of his books. Through the last chapters of the *Structure* he gives us signals to indicate this skepticism. One clue in particular occurs when he refers to himself as an historian. "I do not doubt," he says [preferrable to Henry Kissinger's frequent "I do not think.."]:

> ..that Newton's mechanics improves on Aristotle's and that Einstein's improves on Newton's as instruments of puzzle-solving. But I can see in their accession *no coherent direction of ontological development.* [my italics—DR] (*Structure*, p. 206)

What this passage indicates is that Kuhn is not as much of an historian as he thinks he is. An historian needs a working hypothesis about a "coherent direction of ontological development," not only in physical science but in biological and social science as well. It is tantamount to saying that no one in his right mind should listen to you or read your books if your mind runs helter skelter with respect to historical questions.

"The first (prerevolutionary) and the second (postrevolutionary) paradigms are incommensurable," he says. That is why Kuhn cannot maintain that "it has continuity and convergence toward objective truth." This is the view of A.P. Polikarov in his paper "Some Questions in Connection with Thomas Kuhn's Conception of the Development of Science." (1975). According to Polikarov, the difficulty in Kuhn at this point is that he believes that "..the attitudes of the scientific community are those that are determinative (and not the character of the paradigm—in its content). Kuhn thinks that the major change in the scientific community is a change of *conviction* rather than a change in knowledge based upon an ontological view. The stress for Kuhn, according to Polikarov is not on the scientific cluster of ideas so much as upon the change in attitude of the scientific community. In the parlance of Charles L. Stevenson, he believes more in the change in attitude than in the change in belief. If all the preceding back to "objective truth" is considered, then the transition would not have to be from an old to a new conception, but could be a transition from a new to an old conception—say from Newton to Ptolemy or from Einstein back to Newton. This is an embarrassment for Kuhn because on his view there is no difference between a revolutionary and counter-revolutionary (which is consistent with current U.S.A. foreign policy). It also follows, according to Polikarov that the possibility now rises for Kuhn's position that"..revolution is not affected by its workers, but later on, by the generation of those specialists who adopt, apply and elaborate the revolutionary doctrine." (By analogy, this would seem to fit monopoly capitalist business practice). It might be noted that dynamical problems connected with cosmic flights are being solved by classical means without referring to relativity. Here the revolution is affected to the point of establishing the applicability of the old theory. Furthermore, certain sciences develop "not through a replacement of paradigms" says Polikarov, "but through the borrowing of ready paradigms from already existing sciences." Einstein, for example, did not look upon his contribution as revolutionary, but as a continuation of well-known physical ideas.

Having followed this discussion, we can see that Kuhn is perhaps a chronicler more than an historian since he has no principle of explanation for macro-effects. And this takes us back to his original ontological rejection of the principle of correspondence in accordance with the skepticism of Kant. Without some postulated (at the least) ground of knowability one is hardly in a position to extrapolate the thesis of incommensurability as one moves from paradigm to paradigm in abnormal (revolutionary) science. Yet, he does find resemblances between Aristotle and Einstein. Without a changing infrastructure somehow connected to a superstructure through time this seems difficult. As it is, everything seems to change in the superstructure as far as Kuhn is concerned. And so it is for most idealists. Kuhn is dialectically disengaged when considering development and interrelations of the forces of culture surrounding the scientific enterprise. The power of scientific knowledge—as opposed to religion and magic—is its universality, necessity and accumulatively approaching to objective truth. Religion and magic give a distorted or fantastic picture of reality; and hence Kuhn's pragmatic-positivistic account of the difference between science and non-science is inadequate. This is a diplomatic way of saying that it is false.

A dialectical view of science, such as that put forward first by Engels in *Dialectics of Nature* and later by Bernal in *The Social Function of Science* and *Science in History*[49] proceeds from an evolutionary perspective which emphasizes the accumulative effect of science and its progressive thrust based on objective reality. John Thomspson in his Article entitled "Notes on 'Dialectics of Nature' " notes that:

> The progress in the historiography of science is not just a progress in history *as such*, for without a dialectic between the histories of the sciences and the theory of what contitutes science we would inevitably produce the twin nonsenses of a history that gives equal weight to every historical "fact" and a philosophy whose criteria of scientificity rules out *all* real science as unscientific.[50]

What he here refers to is the tendency in neo-positivism to deal with science as if it were formalized talk about talk about science within a relatively narrow (especially limited to physical science) range. I do not mean to say that physical science represents a narrow range of science in the sense of the number of practitioners or the esteem in which they are held, but in terms of how vast science has to be to encompass nature, biology and social life.

Kuhn is quite inconsistent when he says"..the list of problems solved by science and the precision of individual problem-solutions will grow and grow,[51] "while at the same time he says: "We may have to relinquish the notion, explicit or implicit, that changes of paradigm carry scientists and those who learn from them closer and closer to the truth."[52] Why does he maintain this? If you are solving more substantive problems with greater precision, you must be approaching the truth. Nothing could be more analytic if we mean by truth, correspondence with material reality. If not, then we are remaining the same or getting farther away from the truth. Kuhn tends to be coy here. He cannot say that he doesn't have *any* notion of what the truth is or means because he uses the term "truth" in the phrase"..a permanent fixed scientific truth", and does not at the same time use the escape hatch of pretending that he does not know what he is talking about when he says it. One cannot help notice that he has been at severe pain to avoid using the word "truth" throughout his monograph. Truth is always an embarrassment to the subjectivist because it implies something objective—something that he can't think away, make go away at will as in the popular American song: "Make the World Go Away." Above all, something that doesn't merely depend upon someone's experience of it or the condition of someone's consciousness (mind) is something obstreperous for wishful thinking.

When Kuhn says in the first sentence of his *Copernican Revolution* that "The Copernican Revolution was a revolution of ideas"[53], he is emphasizing that it was the revolution in IDEAS that is most important about revolution in science. If there was any revolution in technology or processes or things (basic to the ideas) which brought about the change in ideas—that is to say an objective change—Kuhn is not much bothered by it. Such events as the influence of navigation on astronomical research, the influence of trade or war upon mathematics and mechanics, the introduction of Hindu notation (the physical spots as well as their symbolic meaning) appear to leave him quite cold. What are the objective, material conditions that made the change in ideas possible? It could be as simple as a bolt moving slightly to the left or glass being distorted. Did Descartes' notion that we needed clear and distinct ideas come from his experiments with optics—with glass? What Polikarov calls the "pragmatic-positivistic view of science" may imply that ideas and their material basis may simply be hypothetical—a continuation of the idealistic separation of hand and brain, material and idea. Despite these important cavils, Polikarov recognised the important of Kuhn's "theory as a disciplinary matrix," in *Methodological Problems of Science* (1983).

Kuhn has a hard enough time believing natural science is a science without being questioned about the status of social science. David De Grood pointed out that Kuhn "unconsciously" discovers dialectic relationships.[54] Hyman R. Cohen, on the other hand, believes that Kuhn is much more dialectical than most positivistic subjectivists (although he does not use the last two terms).[55] I believe, as anyone can tell, that Kuhn doesn't find nearly enough dialectical relationships. If he is the most dialectical of recent bourgeois historians and philosophers of science, he still leaves out serious considerations of biological and social science. Hence he is not a well-rounded dialectician. Without the help of social science Kuhn is not likely to find coherent ontological developments because that can be discovered primarily in terms of social life and historical understanding of the process of knowledge. Compared with Popper, however, Kuhn is refreshingly modest about his own ignorance of social science which includes the dialectical history of social science.

Why has contemporary capitalist society allowed and even encouraged much natural science free from theology—that is to say, objective science? On the view of John Thompson it is "..because 'the bourgeoisie cannot exist without constantly revolutionizing the instruments of production." These instruments of production it has been assumed, primarily hinge on investigations in natural science. There is every reason to think, however, that in the future the heavy emphasis on natural science will be somewhat lessened because a view emphasizing the wholeness of scientific knowledge will be seen as necessary.[56] Thompson points out also that in integrating the sciences (in determiing the scientificity of knowledge), that

> ..which is not directly and necessarily used in the process of controlling nature can be demonstrated by its place within the whole interconnected complex of the interactions, the mutual dependencies of the sciences (which do not negate their autonomy) can only be elucidated by detailed historical inquiry.[57]

We wish to encourage Thompson to get on with this momentous work.

KUHN'S RELATIVISM

Kuhn, like Popper and Lakatos, is a relativist for the same reason that he is a subjectivist. A relativist denies objective knowledge and objective cognition. If one cannot admit that there is "..a coherent direction of ontological development" then one does not see development in history. If he fails to see development in history, then he is forced to believe in conventionalism and what conventionalism implies—relativism and subjectivism. Kuhn, like many another subjectivist who denies that we know objective reality and denies that truth can be known about objective reality, is not quite consistent about this. There are times when he has a favourable attitude towards the possibility of objective knowledge. As De Grood notes: "...while Kuhn's theory appears to be an *historical* view of science, upon closer examination, it does not rise much above psychological institutional study."[58] Kuhn, as we might expect, never states that he is a subjectivist or that he is a skeptic. Nevertheless his remarks indicate that he is both. He raises the question as to whether there is any such goal as that "set by nature in advance."[59] In this he reminds us of Popper's playful [?] personification of nature. Yet he also says: "Can we not account for both science's existence and its success in terms of evolution from the community's state of knowledge at any given time?"[60] Yes, indeed we can. But what does this question of Kuhn's imply ? It implies that Kuhn needs the information available to him from cultural anthropology, historical materialist research, and other social sciences. He writes as if it were not known how to answer his question. It suddenly appears to him as in a unique flash of insight. Yet the information has been readily accessible since the nineteenth century.

Kuhn then proceeds from his position of ignorance of social science to say :

> Does it really help to imagine that there is some full, objective true account of nature and that the proper measure of scientific achievement is the extent to which it brings us closer to that ultimate goal ?[61]

As above, we hasten to assure him that the answer is, yes. It has at the very least heuristic value. Kuhn proposes, however, that since we cannot believe that we are approaching objective truth in science, that "..we learn to substitute evolution-from-that we do know for evolution-toward-what we-wish to-know..'[62] And he adds that if we do this then "..a number of vexing problems may vanish in the process." That is curious! As we give up the belief in objective, public truth, the problems raised because of our need to know what is objective, public truth (an essential of science)—our need vanishes—into language?

Linguistic philosophers believe that the clue to getting rid of ontological and cosmological puzzles is to clear up linguistic confusions. But if we cleared up all the confusions of the language of Plato, Parmenides, Aristotle, Zeno, Leibniz, Descartes, not to mention Heisenberg, de Broglie, Eddington, Jeans and a hundred other natural scientists who opted for idealistic interpretations of the universe, we could do so only with reference to the objective world. Plato's dialogues such as the *Theaetetus* [on epistemology] not only do not clear up confusions but add new, interesting, confusions. And that is not the result of Plato's knowing about as much mathematics as an educated high school student in his freshman year, as Whitehead once observed. Now Kuhn is trying the same stale tack—solving problems by language reform. He labels his substitute for objective truth as an "alternative view of scientific advance."[63] There is no need of an alternative. We have had and we now have objective truth. And if Kuhn will disabuse himself of Kantianism he may have it too.

We surmise that the main "vexing problem" for Kuhn is whether we can know the objective, external, real, material world al all. It seems a pity that his subjectivism and relativism put him in the awkward position of maintaining that science enables us to know more and more relative to man's past history, yet at the same time denying that this means that man is really getting to know more and more about anything "set by nature." This is not merely a psychological problem but an ideological one with roots in his ontology and epistemology. He is the reverse of Tertullian who said: I believe it because it is absurd. For Kuhn it becomes: I can't believe it because it is true. I can't believe it because it is obvious to anyone who has ever tripped on a step, or who, sitting on his chair believed that he was held up by something more than his sensations. (What are his sensations *of* ?)

Kuhn's major difficulties are more easily amenable to correction that Popper's. It was A.J. Ayer who noted Popper's *Umkehrmethode* [not Ayer's word, but DRs]. For example, Popper did not wish to be identified with positivism which is associated with the "verification principle". As Ayer says, the "principle of falsification" which with its interconnection with "verification, "I have called *Umkehr* (turned inside out) seems to be close indeed. It would take us far afield to analyse Popper's contention that the principle of induction cannot be logically justified—but that is a good indication of how flexible Kuhn might be in contrast to Popper who seems to wish to wreck scientific method just so long as he can prove social science (Marxism) unscientific. No principle of science needs to be logically justified (as Popper asserts). Scientific principles need to be empirically and epistemologically justified. Logic provides a formalistic justification in any event—one might say that it is sterile in a sense that Gutzmore intimates but does not openly state. Logic justifies the elements and operations of a closed system. A science, as one might think Popper would be keenly aware, is an open system. Kuhn is obviously not trying any funny business; Ayer demonstrates in his article "Professor Popper's Work in Progress" (*New Statesman*, 1 February 1963) that Popper, on the other hand, is trying to swallow the canary and have it too.

What are Kuhn's major difficulties? First, having no philosophy of history except an unserious one in which he could say that there might well be theological progress even if there were no scientific progress, he cannot understand man's place in nature. Second, he doubts that we can ever know more of the truth of reality or just plain reality through the development of scientific knowledge. And third, he has dogmatically accepted the viewpoint of various positivists, pragmatists or operationalists who have warned him against believing that any teleology can be found in nature, and since for him, nature does not include social man in history (presumably) then there can be no teleology in any part of nature. (Man, as part of nature, provide teleology). His second doubt will disappear with the dogma of Kant that the thing-in-itself can never be known. Kant's position here reminds me of a friendly chat I had with a Professor of French (who was also a dean of his college). He looked at me one day, as we were discussing an old mutual friend, and said: "You know, of course, that so-and-so is older than we think." To know that he was "older than we think" he would have to know that he was older than we once thought, in which instance he is not older than we think. I call this "the older than we think" paradox.

Kuhn might solve a few problems by getting into more hot water. He could join Plato, Popper and Lakatos in a Third

World."[64] This "Third World" is distinct from the first, material world, or the second, world of consciousness. We have already observed his tendency to admit that theories have ontologies, but he believes that these ontologies may reflect *only* mental states or hypothetical entities. Perhaps eventually the real world of nature may prove to be the world through which Popper, Lakatos, Husserlians and Heideggerians annul their subjectivism, relativism and agnosticism. If the "Third World" were material reality, Kuhn's ontological and epistemological problems pretty much vanish. Back in the unified world of nature and consciousness the problems are not always easy to solve, but there is an antecedent probability that they can be dealt with better than ghostly "Third Worlds."

After criticizing Kuhn for his subjectivism, Hyman R. Cohen states that,

> Kuhn has provided a model of the scientific edifice, as it has developed up to the present, which is almost, but not quite, adequate. It is largely dialectical in form and content [?]. It is based on painstaking application of his deep scholarship in the realm of the history of [natural] science, it is certainly more philosophically meaningful than the approach of the metaphysical philosophers.[65]

This statement of Cohen requires considerable comment. First, from the dialectical point of view, Kuhn does connect considerable processes, laws, theories, philosophical judgments as they apply to natural science with some additional reference to scientific institutions (by implication) which are as stated bereft of their political and economic connections and significance. That is his first step towards dialectical analysis. Kuhn is himself a metaphysical physicist. He has much in common with other metaphysical philosophers although it is true that these philosophers have considered themselves to be generally non-metaphysical. Kuhn is a metaphysician neo-Kantian along with other neo-Kantians. He is a metaphysical religious thinker when he suspects that theology progresses. But that is as far as Kuhn goes because he leaves out the role of biological and social science. This alone leaves him without an adequate knowledge of the history and meaning of the term "revolution." According to L.L. Levinson a revolution is a "complete renovation and reopening under new management." Although this is meant to be humorous, it contains within it an essential characteristic of revolution which Kuhn misses—the new management. For example, after the French Revolution the scientific community came under the new management of Napoleon. His paradigm for revolution is bourgeois, not proletarian, revolution under which one may end up with a similar management even if the names have been changed. Kuhn's scientific revolution might have have better called "scientific reform." I think that is what Einstein would have said concerning his own work on relativity, that it was a reform, not a revolution. Levinson's definition of reformer is "One who is trying to make the world a better place to die in."

Despite all this, Kuhn may still be a better dialectician than many of the Western philosophers of science writing from his metaphysical view-point—from the one-sided natural-science analysis of science. Within that framework he has carried forward the problem of the analysis of change. He could in future rely to some extent on the model provided by the Cambridge school of the philosophy of science, including J.D. Bernal, J.B.S. Haldane, Lancelot Hogben and Hyman Levy (see Gary Werskey. *The Visisble College*. (New York, 1979). Praise goes to Kuhn in the development from the earlier *The Copernican Revolution* to the *Structure* to *The Essential Tension*.

As an historian of science and as an American, Kuhn is at some disadvantage *vis a vis* the British science historians. My own hypothesis as to the reason for the relative narrowness of the American outlook as compared with the British is that the British writers on science are given a broader historical and classical training in preparatory school and in the university. perhaps they are less parochial also in their earlier surroundings because they are less isolated from the Continent and other rich cultural milieu. Where they can easily travel to Germany and France, for example, the Americans can travel easily to Canada (but most of them) less easily to Mexico. Canada, despite its many sometimes conspicuous differences from the United States, is nevertheless a kind of client state while Mexico can scarcely be considered a leader in scientific knowledge in the sense that France and Germany are. British scientists generally seem to be more interested in biology and social science and the relationships between society and science in the broader scope (Kuhn emphasizes *scientific* communities). This is noticeable in comparing the work of Kuhn with the following: Bernal, D.S.L. Cardwell, Hogben,Samuel Lilly, Levy, Stephen S. Mason, W.P.D. Wightman, Joseph Needham and so forth. Another possible reason for greater dialectical practice and skill among the top British historians of science is the influence of histodiamat on British intellectual thought because of a less hysterial influence of Cold War ideology in Britain. And if we include the period going back to 1917 the difference would be even more striking.

Most American historians of science still think of "dialectic" as something having to do with Platonic dialogue. And although Marshall Clagett credits V. Gordon Childe in his *Greek Science in Antiquity* (1955) for having been productive in treating of early man, he then scolds Childe for having been "extravagant". The British historian of science, Paul Freedman, in *The Principles of Scientific Research* (1950) makes reference to materialist dialectics, but states in his book that "The author has been advised by his friends to omit the next few pages [referring to the materialist dialectics of Marx and Engels] because they refer to scientific principles enunciated by two men whose names are now associated with bitter political controversies." Freedman had the MS of the book sent to a press in Washington D.C. with the comment that "Such an omission appears to him [Freedman] to be impermissible by standards of scientific honesty. If science, emancipating itself from theology, is to be enmeshed in political prejudice, then it is the duty of all honest scientists to combat such a disaster." (p. 62).

Progress for Kuhn need not be scientific, for he says: "The theologian who articulates dogmas or the philosopher who refines Kantian imperatives contributes to progress."[67] Just as Kuhn does not define science, so he does not define progress. He gives us hints contextually, nevertheless, as to what he means by progress. Progress need not be scientific; it can be theological. Kuhn sees progress of some sort in "..both the list of problems solved by science [and theology as well as Kantian ethics] and the precision of individual problem-solutions....will grow and grow."[68] There can be progress in theology, ethics and scientific problem-solving. There is another consideration says Kuhn:

> In the sciences there need not be progress of any sort, we may, to be more precise have to relinquish the notion, explicit or implicit, that changes of paradigm carry scientists and those who learn from them closer and closer to the truth.[69]

Presumably Kuhn does not believe that the scientific enterprise that solves problems has anything to do with objective reality, but is a gloss on anything enunciated by man (sensible or not sensible?). This may imply that anything that man does at a later time than he did it at an earlier time is "progress". Polikarov, as we saw, noted that Kuhn has made no allowance for a change of paradigm backwards—a reactionary move—say from Copernicus back to Ptolemy if only the scientific community agrees to it. This peculiar position is further clarified when Kuhn notes that the social history of mankind shows no real "process of evolution towards anything." There is no need for speculation about a goal, an end, a desideratum in social process, which clearly contradicts his statement that there can be progress in theology ethics and scientific problem-solving. Can anyone avoid the feeling that this is terribly coy? Some examples, by the way, of reverse scientific revolution include Austrian psychology, German Third Reich sociology, Soviet biology (of Lysenko) and American macro-economics and sociology.

Despite Kuhn's wariness about goals and progress, he nevertheless does again admit something like progress when he says that there is "a..process that move [s] steadily *from* primitive beginnings but *toward* no goal." Here is a clear instance for the need of linguistic philosophy as well as formal logic as well as inductive "logic" and metaphysical thinking—this moving from something to nothing. Why do we speak of "primitive beginnings"? Because we are going on to less primitive states. Why, Kuhn asks himself, as if in an hypnotic trance, "..why should a paradigm change invariably produce an instrument more perfect in any sense then those known before?"[70] "Yes, we echo, "why is that, Professor Kuhn?"

The answer is that Kuhn thinks that the paradigm controls reality instead of reality controlling the paradigm. The paradigm is changed *to fit* reality. Like Milton Friedman, Kuhn believes that science is contemplative rather than an activity of society. In contrast to the contemplative view, let us take Bernal's view of the French Revolution. It was responsible for reformed weights and measures, for establishing the metric system and it created a modern system of scientific education. The first group of students to appear under this new educational system brought about by the revolution included J.A.C Charles,

L.J. Gay Lussac, E.L. Malus and A. Fresnel—al lmen involved in important paradigm change. We need no stick to a narrow interpretation of the impact of paradigm change but move to music, where its wave hit Beethoven and led him to be called "the man who freed music." As great and delightful and as joyous were the quartets and symphonies of Haydn and Mozart, what Beethoven did was to construct a new paradigm for music which was subsequently imitated by many alert musicians. The French Revolution had its effects on Napoleon's attitude towards science and his patronage of it. He often attended sessions of the Académie, took a scientific expedition to Egypt and ordered the Abbè Hauy, the founder of crystallography, to write a textbook on physics.

Kuhn ends his chapter entitled "Progress Through Revolution", by admitting that:

> The world of which that community is a part must also possess quite special characteristics, and we are no closer than we were at the start to knowing what these must be. That problem—What must the world be like in order that man know it ?—was not, however, created by this essay.

What a way to begin! We decide what the world must be like if we are to understand it. What if we first of all asked ourselves: What must a girl be like before we are able to marry her? instead of finding a girl and marrying her. There are succinct descriptions of such an unproductive approach in working-class circles which require a succession of expletives to adequately express scorn for such diddly-dadburned hoity toityness. It rouses the emotions like Popper's blinkety-blanked picyunishness. What kind of people are the dad-blamed idealists? It is the contemplative view of the sputtering lamp which Plato, Berkeley and Kant foisted upon mankind. Why this agnosticism that undermines the bravely fluttering soul of mankind. Suppose that the directors of the sputnik or moon-shot space programs had asked: What would space have to be like before we could contemplate knowing it? They would be caught up in endless epistemological and linguistic debates. I can't imagine what Kuhn thinks that scientists know when they know something. My curiosity is aroused by such aberrations. After all that man has done through science since the nineteenth century can one stand or sit there and doubt that he doesn't know about nature or reality? Surely this verges on hysterical mysticism.

While it is possible to imagine that such subjectivist philosophy of science has made contributions to the history and philosophy of science, these have been mainly in micro-studies which are useful to more comprehensive accounts. The work of scholars like Popper, Lakatos and Kuhn must be judged for what they are: obviously partisan, skeptical, negative, metaphysical (this is a matter of degree among them) and formalistic.

PROSPECTS FOR THE FUTURE IN THE HISTORY AND PHILOSOPHY OF SCIENCE

One wonders how much of the subjectivists' writing will endure for a significant length of time. There does seem to be a significant body of opinion that what Kuhn has done is significant. It can be incorporated into the body of philosophy of science initiated by Engels and advanced by Bernal. What is crucial to advancement now is that the whole field be cleansed of its one-sided, parochial, formalistic, subjectivistic unhistorical perspective. We need many more scientists and philosophers as well as historians with a comprehensive dialectical outlook embracing not only the natural and physical science which is the topic of most philosophy of science for a hundred years but also the biological and social. The surface of interconnections among these has barely been scratched.

The social sciences from anthropology to sociology should be seen in their interrelationships to social history with constant reference to advancement in physical and biological science and technology. The tendency to treat technology as a foster-child should also cease. This requires an international or global perspective since no single scholar in a single country can carry out such work by himself. Co-operation is mandatory. We have co-operative models before us such as Needham, Langevin, Joliet, Bernal, Einstein, Oparin, Varga, Kapitsa, Struik, Debibrasad, Rahman, Infeld, and in his last phase, Bertrand Russell.

Social scientists should take a broader perspective which puts more emphasis not only on the infrastructure but world trade and the winds of ideology. We need studies to indicate how things got done and do get done despite or in addition to a thousand ideologies and misinterpretations circulating throughout the international order. There are more than 150 countries each of which has its slightly different interpretations of reality.

Indian intellectuals involved in the history and philosophy of science should look for more realistic interpretations of reality, pay less attention to the interesting but counter-productive subjectivist explanations offered by Western European and North American specialists. Asians, Africans and Central Europeans, thinkers in the South Pacific and the Caribbean have additions to make as have those in Latin America. Indian philosophers and scientists require complementary counsel. Social science should be studied as well as physical science helping to weed out obscurantist and apologetic material. Critical cultural anthropology, critical sociology, critical epistemology concerned with objective reality rather than with more formalism are much needed. Indians have shown themselved through the past historical epochs to be keenly conscious of mathematics, logic and formalization. The future of Indian work in the philosophy and history of science depends upon a commitment to realistic and materialistic studies within an historical framework. This is what such pioneers as Debiprasad and Abdur Rahman have been heroically trying to do in their promulgation of the science of science for the past thirty years,

IV

SUBJECTIVIST ECONOMICS IN THE FIGHT AGAINST OBJECTIVE SOCIAL SCIENCE

INTRODUCTION : ORIGINS OF MODERN SUBJECTIVISM

Modern subjectivism begins with Kant (1724-1804) whose role it was to admit the importance of scientific knowledge while at the same time undermining its foundations. Claiming that Hume had awakened him from his dogmatic slumbers Kant "awakened. . [and descended] the other slope of the dream."[1] The thrust of science throughout the world has been materialistic, whereas its major antagonist, has been the conservative espousing of idealistic thought, nearly always the defender of fideism. In order for Kant to achieve his end, that is to make room for God, immortality, and freedom as understood in Christian theology, it was necessary for him to attack scientific or materialistic ontology and indirectly scientific method. He did this by maintaining that the real independently existing material world could never be known because of the inhibitions of human nature. It was human nature to know only appearance, sensation, and not the material objects giving rise to this appearance. This is obviously contrary to the requirement of science that we generally know the work of appearance to be what it appears to be, giving us knowledge of the external material world. Kant's view is reactionary in this respect[2]. Having thus limited human knowledge to appearances of something we can never know, which is in itself logically as well as ontologically absurd, if we are to be good followers of Kant, we must act in such a way as to pretend that the world of appearance is the world of reality. This gives priority to sensations and consciousness as well as to the fideism which feeds on them. On the contrary, science gives initial priority to matter—to the independent objective world. Science protects us from subjectivism by enabling us to base our knowledge on the ontologically real despite J.L. Austin's red herring. It was his contention that "real" is used merely "to exclude possible ways of being *not* real." But why bother to do that unless one needs to know the ontologically real ? At the same time, so that we yet maintain the appearance of science, as it would be unthinkable by the eighteenth century not to permit at least its image, we must have faith because this is necessary for moral order, since on the idealist view moral order cannot be based upon material reality. That Kant gives no satisfactory proof of this has not deterred many, Western natural scientists and social scientists from accepting it. Kantianism was also readily accepted in Japan, China, and India, which, as semi-feudalistic countries in the early twentieth century were strongly in the grip of idealism already.

While promising to give us a unified philosophy, Kant ends up destroying the possibility of objective knowledge of the real material world which he replaces by sheer faith to the satisfaction of idealists and fideists. In the meantime physical and biological science continued to advance by simply acting as if Kant were mistaken. Social science, however, despite a strong beginning in the eighteenth century, seems to have been ambushed by Kantianism because that philosophy proved so useful a tool to prevent a serious critique of social life under feudal and capitalist institutions.

After Kant, a member of the Baden School of Neo-Kantianism, Heinrich Rickert (1863-1936), provided the following conceptions as glosses on the postulates of Kant. The most important of these were :

(1) All that exists to us is evolving consciousness.
(2) Man creates a transcendental world of thought, but there is also a real world of unknowable objects "behind" it.
(3) Truth is the building up of a reality of thought (transcendental world).

These are the tenets accepted by most modern subjectivists.[3] Taken together, they imply that objective science is impossible and that we are totally dependent upon our consciousness, for whatever knowledge we possess. Marvin Farber has correctly noted that "The central theme of the subjectivist is the analysis of experience."[4] Hence the subjectivist soon finds himself examining the content, not of the experience *of something*, but the content of the experience *of experience*. If, on the other hand, we are going to accept an objective point of view, we admit the fact that consciousness and experience take place within a larger sphere, namely in objective nature which exists independently of us. Of course we can make certain changes and alterations in nature which are demonstrated in the history of technology. Everything that happens in the universe, according to the believer in objectivity, occurs within the framework of physical nature.[5] To believe that physical nature is merely an experience, a happening, within our consciousness, is to be an idealist and subjectivist. The consequence of this idealistic position is that whoever holds it rejects the objectivity of science and believes that everything we know is subjective or appearance only.

Why was this subjectivism accepted with barely a shot fired, or so it seems from our perspective today ? The reasons, I think, have already been given. First, materialistic science threatened the world-view of every idealist[6] because it implied that everything in principle could be objectively known. If it could be objectively known what about religious, political, and social beliefs ? Couldn't their precarious position be subverted by objective inquiry ? The answer was learned in the eighteenth century before, during, and after, the French Revolution. Yes, objective inquiry can potentially subvert everything that is false. But what is false may be dear to many hearts. Therefore, we must fight against such inquiry, say the subjectivists. To fight against objectivity, one must fight against objective materialistic science. Consequently, the fight against science, materialism, and objective inquiry, has been a major concern of most institutions and individuals armed with the false postulates of Kant and Rickert. The armament, as we shall see, became fairly complicated in the ideological struggle between idealism and materialism, subjectivism and objectivity, anti-science and science. Man's ingenuity in the arts and crafts is repeated in his ideological lapidary work in economics, philosophy, and sociology. It is well worth reading if one has the time. Somerset Maugham relished studying it.

Subjectivism is spiritual apologetics. If one wants to learn about the objective world so that it can be controlled, be improved for human beings, he does not indulge in apologetics but in science. Subjectivism is not only the opposite of science, but is the spearhead of the fight against science. Scientific socialism is science applied to society, so it also must be fought against, according to the practice of the subjectivists. Science, however, has become a powerful tool and benefactor over the centuries, so that only a madman will now attack it directly. Hence, subjectivism covers its activities by pseudo-scientific jargon or a mystifying technical apparatus to hide the fact that it is subversive of the greatest single development of man's ingenuity—science. It is no-accident that contemporary masters of yoga call their art "The Science of Yoga," or that Mary Baker Eddy, called her eclectic religion "Christian Science," or that Bertrand Russell called his philosophical subjectivism "Scientific Philosophy." Yet all of them reject the essence of science in refusing to recognize the objectivity of science resting upon a knowable material world independently existing with or without consciousness. Russell may have finally admitted this in his nineties. If all primates, including man, were destroyed, there would still be a material world doing what it is necessary for it to do.

SUBJECTIVISM AND OBJECTIVITY IN EARLY SOCIAL SCIENCE

The proto-social sciences were magic and ritual. Within them they carried the seeds of economics, sociology, philosophy, and the other social sciences. Economics was the first social science to develop institutionally during the eighteenth century, imitating as best it could the outlook and techniques of physical science. The religious, national, and class struggles which accompanied the growth of capitalism, led to a new inquiry as to the ultimate purposes of society. The criticism of Lilburne and Winstanley, the Levellers and Diggers, was the earliest to combine theory with practical reform which culminated in the work of Hobbes (1588-1679).[7] It was Sir William Petty, once secretary to Hobbes, who began economic statistics which he summarized in his *Political Arithmetic* (London, 1690). Petty lays down in this book the labor theory of value which was directed against courtiers, churchmen and landlords who claimed rewards for which they had not worked.[8] As Bernal says : the labor theory of value ". . . . fell into disrepute among orthodox economics because it was felt that it might be used in the same way against capitalists."[9] And so it was, and so it still is. Apologists for subjectivism replace it by the mysterious "marginal utility" hypothesis.

The notion of surplus value, by which Marx explained the difference between the value of a product (good, commodity) and the wages paid to produce it, is implicit in Adam Smith (1723-1790). Smith provided a justification for the practice of industrial capitalism and the attitude of "enlightened self-interest." Soon thereafter the Physiocrats appeared in France with their catchwords of *laissez-faire* and *laissez-passer*. They mean essentially "leave us alone to conduct our business and sales." Comparative sociology, based on the reports of travelers and missionaries in China and India, appeared with Voltaire (1694-1778) and Montesquieu (1689-1755). This new science was used to fight the relics of the feudal order. Being unable to appeal to tradition, the sociologists (*philosophes*) justified their criticism on the basis of a "hypothetical rule of reason."[10] The French Revolution emphasized in its first phase the rights of man, and in the next, the rights of private property . Babeuf (1760-1797) made the fatal error of calling for economic as well as political rights and was consequently executed. After the fall of Napoleon, economics was essentially the justification and apology for rising industrial capitalism. Yet there was famine, pestilence, ignorance, and social insecurity in the heart of England, then the wealthiest country in the world. All this had to be glossed over or made to appear to disappear. It was up to religion, economics, sociology, and philosophy to help in this. Comte, for example, stressed that "......Materialism is a danger inherent in the mode in which the scientific studies necessary as a preparation for Positivism were pursued."[11] He also recommended the introduction of a Religion of Humanity which will bring "Feeling, Reason, and Activity into permanent harmony,"[12] all under the "direction of the Heart."

Radicals also appeared. Robert Owen, one of them, was described by an admirer as a man "who made a fortune by capitalism and common sense and squandered it in communism and craziness."[13] Owen tried to help workers through a co-operative moment. In France, Saint-Simon and Fourier thought that socialism would bring social justice into being and as soon as that was understood, they believed, no one would hesitate to recommend its acceptance. Marx borrowed from French socialists, British political economists, and German classical philosophers and made socialism a science since he could explain the inner dynamic of many social movements from ancient times to the future. This dynamic is the struggle of oppressed but rising classes who appear because of technical and economic changes in methods of production. But in order to encourage class struggle in getting rid of capitalism, Marx emphasized the need for class consciousness, a realistic assessment of one's place in society. As we have already noted, subjectivism fights against materialism, science, and socialism. But it fights against these with special ferocity when they appear in their most unified and unpalatable form—as histodiamat.[14] Subjectivism's loathing for this position is complicated by its elite-ism (demonstrated in its transcendentalism and spectator-contemplative view of knowledge, knowledge which consists of knowing without doing (producing). Its hostility has led it to not only attack scientific socialism, but also to actually defend capitalism without regard for the historical facts.

Orthodox social science throughout the world was unified through nothing so much as by its opposition to histodiamat. At the same time its subjectivism constantly militated against its attempt to appear to be scientific. No amount of juggling with mathematics, no effort to increase the use of statistics, ever completely hid its anti-scientific bias.[15] Since subjectivism above all tried to de-ontologize the world, it is impossible for it to find objective knowledge. Objective knowledge, to the embarrassment of subjectivism, must be based on knowledge of *objects*, not simply knowledge of the experience of *experience*. Histodiamat comes into this as the incomparable target, for as Debiprasad Chattopadhyaya has pointed out : "[histodiamat] is the only philosophy the [truth of] ....which....is already verified by human history."[16] Subjectivist philosphers, for example, are outraged because whereas under socialism advice is solicited of philosophers who are considered to be scientists, under capitalism they "....are left to live the life of queer creatures seeking delight in whatever they choose, not excluding booming nonsense, reckless eclecticism and insipid scholasticism, not to speak of course of such banalities as sheer careerism and job-hunting."[17] There is little doubt that the philosophers under capitalism are the chief clowns of the social sciences.

Philosophical diremption from reality has encouraged various merry-go-round and circus schools from linguistic analysis to transcendental phenomenology. In addition to this, some of the subjectivists are aware that working scientists under socialism are given a sound training in social science, including philosophy, whereas under capitalism the only scientist or engineer to be found in a philosophy class or seminar got there by making a wrong turn on his way to the cafeteria.

GROWTH OF ORTHODOX ECONOMICS AND ITS PARTICULAR FORMS OF SUBJECTIVISM

By the nineteenth century the growth of economics was demanded, not only because of problems of production, but also because of problems of distribution. These two major clusters of problems, of course, as Marx points out in *Capital*, can be only artificially separated. After 1825, the series of never-ending crises of capitalism[18] required careful scrutiny and explanation. This was accomplished by Marx and Engels who based their work upon certain premises of classical economics, a wide knowledge of ancient, medieval, and modern (including Asian) history, keen awareness of industrial practice and physical science, a knowledge of mathematics, sympathy for the plight of the worker who produced the commodities the capitalists got rich selling, the technical equipment of knowledge of the major European languages, and sufficient physical and mental energy to work unstintingly.

Classical economics was essentially macro-oriented and objective, being a description of what actually occurred in industrial production. After Marx and Engels, however, orthodox economics becomes increasingly micro-oriented and subjective, its major thrust being to defend capitalism by advancing pseudo-theory (when it had the ingenuity to construct it) or by eliminating real-theory which was likely to point out the internal contradictions of capitalism such as were leading to horrendous social failures like mass unemployment, arms production and war. The micro-approach was also the result of the greater specialization (splinterization) which follows from capitalist development both in the productive base and in the superstructure.

It was economic subjectivism that led the way for the other forms of subjectivism in philosophy, history, sociology, anthropology, psychology, law, and linguistics. It also appeared in physics, but that does not directly concern us here.[19] Anthropology, when it first appeared, like economics, tended to be objective and scientific, especially under L.H. Morgan (1818-1881) and E.B. Tylor (1832-1917). Marx and Engels used the material and categories of Morgan to demonstrate the common social structure of tribes in different parts of the world, the common tracing of relations through the mother, the absence of tribal property, prisons, or police. Although some of the anthropologists carried on similar work in this century, for example, Robert Briffault, Benjamin Farrington, Bernhard Stern, and Leslie White, it was in general believed to be safer to do as Sir James Frazer or Westermarck did : collect art objects (masks), skulls, bones and record folk-lore (and later their successors, folk-music).[20] Their early museums resembled magpie ness.

In economics, micro-considerations and mathematization increased because of the needs of capitalist bookkeeping, the major interest of which is accurate records[21] of profits and losses, and wherever possible, prediction of profit. Statistical methods have increased geometrically since the last quarter of the nineteenth century, but it must be remembered that by themselves, these do not contribute objective information. The postulates, presuppositions, theorems, and equations, whether related to reality or simply to subjective projections will determine the relevance of quantitative studies.

It is no accident that the Austrians and Germans, who were most distant from the degradation of capitalism, should be the most sanguine about its future viability.[22] They also knew the least about anthropology, despite the appearance of a few Humboldts. From the economic heart outwards, attack on the objectivity of science and in particular histodiamat, usually radiating from Vienna, has been a constant thrust of orthodox economics.[23] Orthodox economist reaction to histodiamat has varied from mere hostility to hysterical polemic. By "histodiamat" we mean the combination of historical and dialectical materialism. (Whereas Marx said he was not a certain brand of French "Marxist," he never said he was not a protagonist of histodiamat). While pretending to be neutral and unemotionnal about economic science, the orthodox economists like Walras presented a strong case for *laissez-faire* against the followers of Saint-Simon, or like Boehm-Bawerk who much as Karl Popper, wanted above all to destroy the influence of Marx. As Eric Roll points out :

> The existence of this conflict between the conservatives and the radical interpretation of classical doctrines was well calculated to make economists uncomfortable. Marx increased the discomfort by carrying classifical doctrine to one extreme though distorted conclusion. To make economists face up to these classical contradictions resulted in irritating them. The outcome of this irritation was often to abandon objective judgment.[24]

As we shall discover, there was no need for them to "abandon" objective judgement because by their presuppositions that the real material world cannot be known, they never had any. Having abandoned the possibility of objectivity they are left in the slough of subjectivity where they perform the bourgeois rot.

The Austrians spearheaded the counterattack on histodiamat. One reason for this is that Austria, although the least capitalistically developed of the more important Western European states and which has served (for a century) as a center for orthodox and reactionary ideology, had the most influential university in Europe at the end of the nineteenth century. Another was that histodiamat, having its roots in economics objectively based upon the most advanced practical knowledge of capitalism, was simply more sophisticated from the beginning. The Austrians with their German Allies, switched from the advanced classical approach [which emphasized objective production, since that was what England and Scotland were concerned with] to an emphasis upon consumption, demand, and utility—concerns of the more clientistic states of Europe. The Austrian reactionary view of the world changed emphasis upon production to consideration for "rentier economics." Neutralism also developed, implying the separation of ethical questions from questions concerning the market. The classical labor theory of value was replaced by the marginal utility view, a serious constriction dictated by the need to avoid admission of the deprivation and exploitation rising,[25] first from *laissez-faire* capitalism, and then from monopoly. Even the nearmyth of government intervention does not hide the nature of capitalist-labor relations, nor the increased need for exploitation. What made the labor theory of value such a nuisance is that it "has the merit of directing attention beyond questions of 'explaining actual prices' to questions about social labor and the appropriation of its products which are ignored by those who simply want to study price fluctuations."[26]

The alteration from objectivity to subjectivism by underscoring consumption changed classical economics to an atomistic, unhistorical, and narrowing psychological study. Since foundations, except formalism, must be ontologically objective, this means that orthodox economics did not really have a foundation, which should have been a source of embarrassment to its practitioners ever since. Instead of the historical view of the structure and physiology of society as a whole, society is narrowed to appear as an agglomeration of individuals. The productive base is replaced by a psychological "substructure." Just as it is perverse to claim that the trees hold up the earth, so it is vexatious to maintain, as the orthodox economists do, that the superstructure holds up the base. The economist now replaces the political economist by "stating the obvious in terms of the incomprehensible," as Alfred A. Knopf put it.[27] There is some truth to the statement that orthodox economics is a branch of psychology, because economists, instead of trying to predict economic events seem to be content to try to predict psychological ones.[28] Many of them constantly stress such things as "the mental climate" of investment and "the mood" of the public instead of producing useful objective economic data. Excesses in bourgeois psychologizing probably reaches its height in political science. In his first week in office, President Valery Giscard d'Estaing is said to have brought "a more relaxed style to the presidency" of France. This "relaxed style," mentioned as a sop to Giscard's disappointed and dangerous opponents, presumably consisted in "moving friends into key positions," walking rather than riding in cenral Paris, and promising journalistic freedom even to those "who attack me personally."[29]

Subjectivism of a type was present in the West before Plato, but much of contemporary and modern subjectivism starts with Kant because whereas Plato was simply anti-materialistic, Kant was also anti-scientific in order to save fideism hence the complexity and tortuousness of his arguments for subjectivism.[30] Kant's subjectivism is mostly induced by theological considerations; the fideism in Neo-Kantianism is patent as can be seen in the writings of Lotze, Windleband, Rickert, Brentano, Husserl, and Heidegger. Rickert contributed to the classics of subjectivism through his formulation of the following tenets which we repeat here and then compare with those of histodiamat to bring out the difference between anti-scientific and scientific consciousness.

> According to Rickert: (1) All that exists to us is evolving consciousness. (2) Man creates a transcendental world of thought "behind" which is a real world of unknowable objects. (3) Truth is the building up of a reality of thought in the transcendental world which implies that the test of truth is subjective, being inner certainty.
>
> Contrast these with the comparable tenents of histodiamat: (1) All that exists is material reality in its various forms including consciousness, one of the last to develop. (2) Man learns about knowable and changeable material reality by means of his senses and intellect working together with objective criteria learned through social practice. (3) Truth is made up of all the statements about reality that correspond to it. (4) The test of truth is scientific, social, public practice.

The Austrian and German social scientists depended heavily upon these Rickertian postulates. They were later tacitly accepted by the Italian, French, Dutch, Scandanavian, and Anglo-American, social scientists. Not only was the study of history generally neglected, but also the study of the history of philosophy. In neither study did objective knowledge appear as the result of examining the most general thoughts of mankind. The subjectivists, like Wilhelm Windleband, found that "the history of philosophy presents to our view the gradual attainment of clearness and certainty respecting those values whose universal validity forms the problem and field of philosophy itself."[31] The encyclopaedic Friedrich Ueberweg, one of the earliest subjectivists to look at philosophy piecemeal, has no conclusion or general summary at all at the end of his two volume work.[32] B.A.G. Fuller ends his *History of Philosophy* with a typical Anglo-American flippancy by saying: "These things lie on the knees of the gods,"[33] presumably because the lap would be too sexually significant. Julián Marias finds that:

> The history of philosophy comes to a close in the present, but the present which is laden with the entire past, bears the future with-

> in itself; the mission of the present consists of setting the future in motion.[34]

What we learn from this is what we learn from the critic who says that a sentence has a beginning, a middle, and an ending. It is only when we turn to openly polemical history that we find serious conclusions drawn. For example. V.V. Zenkovsky concludes his two volume history of Russian philosophy by "..having discerned in the development of Russian philosophy the basic importance of themes of secularism [versus Christian ones.. in which] the genuinely vital creative problems of philosophic reflection derived from the Christian gospel and hence cannot be resolved apart from it."[35] What Zenkovsky says may not be objectively true, but he at least does not hide behind the fasade of neutralism or pseudo-objectivity.

Meanwhile, the Austrian and German economists had use sedilla agreed that since reality cannot be known, uncertainty, inertia, and indifference rise in importance. They are necessary to an understanding of what is left of the economic body after the heart, liver, and lungs have been removed. Uncertainty rises when there is no ontological base to which to refer substantive issues; inertia rises because it is difficult to see in which direction to move if the relationship of the productive base to the superstructure is a mystery; and indifference occurs because one is talking about subjective, unmeasurable—even unidentifiable—factors. Indifference acquires the mantle of symbolic merit as in Pareto's indifference curves. It was one step from the indifferent to indifferentism, an example of which is the dissociation of economics from social or political consequences. Uncertainty is associated with uncertainty of profits—in the broadest sense—the lowering rate of profit* which hangs over capitalist apologetics like a cloud. As long as economics was concerned with *production*, the uncertainty had to do with capital, material, and labor. But under subjectification, economics spends its time worrying about buying and consuming. The *New York Times* letters to the editor show deep concern under the Presidency of Reagan about U.S.A. decline in *production*. Everyone wants to make money without producing anything. Profits, for example, became subjectively mystified by the American economist Frank H. Knight, who says that they are the "result of ignorance of the future." His cowleague, Milton Friedman,

*Marx predicted a falling rate of profit for the capitalist system as a whole over time.

adopts this also, referring to profits as a "result of uncertainty."[36] In other words, profits are the results, not of capitalists appropriating surplus value from the worker, but of a state of mind. We recall that in Heisenberg too, uncertainty is not in the material world but in the experimentor's consciousness. Marriage and war are also often the result of uncertainty. Sometime profits are the result of near-certainty, when someone has the inside dope on a secret transaction. But generally this kind of reasoning is unscientific and not to be taken seriously. Inertia results in one's being unable to decide when to sell or at what price. It doggedly follows bourgeois economists' ignorance of reality.[37] They must be content with trying to learn about states of mind." It is the state of the consumer mind, not that of the objective world, that is important to subjectivists.

After the "laws" of H.H. Gossen (1810-1858) became generally known, subjectivity in economics was given a suitably specious foundation. Gossen's "laws": satiety,[38] maximization and curtailed supply may be stated as follows: (1) satisfaction decreases with each additional good—this is the basis of marginality or marginal utility which replaces the labor theory of value. (Each additional billion seems to heighten the satisfaction expressed by billionaires such as J. Paul Getty, who became more euphoric with age.) Otherwise, from whence comes the descriptive phrase" "insatiable greed"? The first law of Gossen requires basic revision and perhaps endless qualification. The labor theory of value, we should be reminded, was intolerable to the orthodox after the socialist critique of capitalism by Marx and Engels, since it revealed without camouflage that the capitalist was an exploiter. As Nehru said in the *New York Times Magazine* 7 September 1958; "Sociali:m is, after all, not only a way of life, but a certain scientific approach to social and economic problems." Furthermore, it was proven that if the capitalist were not an exploiter he would soon disappear from his privileged position, and hence, regardless of his own volition was determined by the system to act as an exploiter. (2) Demand keeps the marginal utility of goods equal. The fascination of orthodox economists for equilibrium is a study in itself. It can be partially explained by the fact that being idealists the orthodox have a tendency to approve whatever is basically changeless. In addition to that is the alleged "balance" required to make the obviously anarchistic capitalistic system appear orderly. In recent years "countervailing powers" were raised up to new importance by J.K. Galbraith who lately seems luckily to have forgotten about it. (3) Value attaches to a good only when its supply is less than the demand. From this it follows *mirabile dictu* that any commodity which one already possesses has no value which is why it is insured! This is the origin of that surrogate required from the time of Gossen: sentimental value. Extending this thought to the consciousness of capitalists themselves, it is not what one possesses *in situ* that is important, but what one can possess *in potentia*. Who could take this seriously?

The profit that one expects or hopes for in the future depends upon "ignorance" according to Knight and Friedman. To have profit rise from "ignorance" rather than from exploitation makes the capitalist and his idealogue feel more comfortable. In this sense alone, the capitalist is extraordinarily comfortable despite the weight of all the "valueless" goods with which he surrounds himself. The ubiquitous explanatory power of "human nature" is useful here in covering-up the objectively ascertainable exploitation. Man's "inherent" ignorance of the future is a constantly useful ploy when bankers must explain to elderly widows why their savings are no longer available for withdrawal or why they lent billions of dollars without proper collateral to virtually bankrupt countries.

The underlying presupposition of these three "laws" of Gossen is that the aim of human conduct is the maximization of pleasure.[39] Sigmund Freud, the Viennese psychologist, accepts this as did most nineteenth and twentieth century subjectivists. Consumerism adopts it today and it is the psychological basis of the advertising business. As much money is spent on advertising as on education in the U.S.A. In recently flying from Madrid to Frankfurt I read in a German magazine about "*nützlich* Tips" or tricky ways to sell products that nobody needs. In computing the Gross National Product (which has a double meaning) North Americans include advertising revenue which might be better subtracted from it. Advertising is justified on the grounds that it is "consumer education." North Americans now advertise their academic courses in newspapers: "Sex and Transvestism in Renaissance Philosophy."— Philosophy 303. "Sex, Science and Sin"; "Heidegger, De-ontologizing and Smoking."

Subjectivists in economics as well as the other social sciences adhere to the doctrine of timelessness[40] such that value-considerations are operable and analyzable in *any* social order. If they can operate in *any* order they can operate in *no* order, which is precisely where subjectivists require that their terms be employed. The labor theory of value, the foundation of classical and histodiamat economies, is objective, quantitative, and mathematizable. It is replaced by the marginal utility theory which is subjective, qualitative and psychological. Marginalism is the subjective "base" upon which economic thought rests subsequent to the appearance of the "first generation of marginalists," namely, Jevons, Menger, and Walras.

The marginal utility subjectivist theory of value implies that value comes from future use (plus ignorance) of a good in consumption. As we have remarked earlier this odd view implies that neither past nor present goods have any value. This in itself should have been a warning flare as to how unrealistic the orthodox view was becoming. Classical economics and histodiamat claimed that the value of any good, past, present, or future is determined by the labor power (or social labor) expended on it. Consideration of time becomes extremely important because of this orthodox economics problem. Time is no problem under socialism since the labor theory of value accepts commodities as valuable at any time. "Futures," hedging in futures," are analgesics to lower the pain of uncertainty. We begin to hear about reversible time, subjective time (introduced by Kant), eternal time (introduced by Parmenides), time abstracted from temporal duration, social time, biological time, evolutionary time, transcendental time, relative time. We shall discuss this in a later section.

Marginal utility (*Grenznutz*, lit. boundary/limit+useful/profitable) determines the limit of profitability for the seller, that is to say, the capitalist. It concerns the "final degree of utility," for Jevons, "rarity" for Walras, and "least actual utility" for von Wieser. This is a subjective determination such that an individual determines subjectively at any given time how much he desires a given article, commodity, good, utility. The article's worth is then judged by its desirability instead of its useful or embodied labor power as in the labor theory of value. Marginal utility it should be noted, does not exist in the real material world nor has it ever existed in the real material world like the use of muscle power in the labor theory of value. Marginal utilities exist only in the minds of men. Hence, they cannot be transmitted between individuals except subjectively.[41] It is no more of a science than the *I-Ching* and probably more dangerous. Since objective causation cannot occur for the subjectivist economist, marginalist causation, subjective causation, is called "imputation". "Imputation" does not provide for any necessary connection between or among objects, however. From this it follows that "imputation" is simply a fiction by which subjectivist, marginalist economists attempt to tie-up subjective entities to give an appearance of science. Since science demands objective connections, present day orthodox economics can never be a science and has never been a science.

In order to measure the unmeasurable marginal utility factor, von Wieser had to develop a number system not connected to the decimal system because the decimal system is based upon objective reality (originally on the ten fingers). Vilfredo Pareto went one step further. He claimed that not only is marginal utility unmeasurable, but that it is unidentifiable as well.[42] It is fair to ask, then, if marginal utility is an unmeasurable, unidentifiable, fiction, how can decision-making be based on it? The answer is that capitalists involved in making economic decisions ignore this fiction and instead use rule-of-the thumb *ad hoc* techniques which are sometimes based on reality. When it was mentioned at a doctoral examination that the candidate was talking nonsense, A.N. Whitehead noted brightly: "Yes, but is it the right kind of nonsense?" (Thus indicating himself to be a pluralistic bourgeois philosopher). "Marginal utility." "presuppositionlessness" in phenomenology, "libido" in psychoanalysis, and "transcendental solipsism" are not the right kind of nonsense. When relatively bright orthodox economists are asked to explain marginal utility they look rather embarrassed if they have any sensitivity. One said to me: "Rather than define or describe marginal utility, I would rather go out and polish my radiator." No amount of polishing will brighten up this by now hoary piece of orthodox dogma.

Before 1854 objective production was the major problem of capitalism and its idealogues; after the eighteen seventies subjective cost became the major problem. Emphasis on *making* was replaced by stress on *selling*. In terms of subjectivist philosophy, the knowledge of changing objective reality is displaced by the subjective knowledge of our own states of mind. According to Alfred Schutz, Scheler had the happy turn of phrase to describe it as "absolute personal privacy" when we finally push our self-awareness to its ultimate.[43] It is this same Schutz who finds that the external object "is constituted out of appearances."[44] Is he related to the view of Nelson Goodman who wrote *The Structure of Appearance*?

A.K. Coomaraswamy has shown the spinoff of subjectivism in the arts where exchange value, the monetary selling value of art, replaces the essential features of an art work, namely its attractive use-value. Making things well is replaced by selling them fast. If capitalists could move at the speed of light (with their idealogue satellites), think how many additional turnovers of capital they could implement in twenty-four hours! But more of this later. One is reminded of André Gide's sentence that ". . . . the mind stifles in this atmosphere of organized falsehood."[45] We now turn to a discussion of what scientific methodology is and what it cannot be for subjectivism.

A CASE STUDY OF SUBJECTIVIST SCIENTIFIC METHODOLOGY

We have been skirting the impossible in trying to explain orthodox economics. Now we shall accomplish the nearly unobtainable by describing one case of orthodox scientific methodology. Since the subjectivist (orthodox social science) has no knowable external material world, he can have no science. But we cannot leave him without analyzing the aberrations involved in his pretending to do science and simulating scientific methodology. His ingenious methods are just as interesting as the methods capitalists use to cover the fact that they live high on the hog by exploiting other people or that the pious use to hide the fact that they are fascinated by sex.

But before describing an example of subjectivist notions of science and scientific method, I wish to state what I believe science and scientific method to be. Science is a representation of objective reality constantly improved upon by social practice throughout history. As Paul Crosser puts it: "All science and all scientific propositions are forms of approximation of reality."[46] Their truth is verified and constantly made more precise. Throughout history scientific description of reality comes closer and closer to the ideal of absolute truth where the probability of the truth of propositions is 1/1. Man, at any given time has through science relative truth about objective reality. Science is important because it finds and studies objective laws upon which truth is based. The importance of science has increased enormously in the past hundred years such that now any major productive system in the world is nearly inconceivable without it. Subjectivism, antithetical to science, sees reality, not as an objective material process, but as a psychological experience based upon individual reactions. Experience replaces reality for subjectivists. Experience and reality have close ties, of course. Instead of recognizing the external objective source of much or most experience, subjectivists see experience as the source of all knowledge. The ultimate ground of subjectivist epistemology is the experience of experience. Empiricism, which holds experience to be the source of knowledge, has two branches: (1) passive empiricism, in which the human consciousness is a passive instrument receiving experience and (2) active empiricism, in which human consciousness and thinking play an active part in receiving experience. The second type represents the more advanced mode of subjective idealism since it is an analog of histodiamat. Orthodox economists usually write as if they primarily believed in the first type of empiricism, but a few of them admit to the second type when they are called to account.

To know what an orthodox economist is talking about, we should check his ontology because that is just as fundamental as his methodology. The two cannot be separated. That they are separated is of course a result of the proliferation of specializations that occurs under bourgeois intellectual life. What then about the orthodox economist assumptions—metaphysical or dialectical? Milton Friedman, a well-known American economist, can serve as a fair sample of subjectivist "scientific" method. He has been warmly acclaimed in both Pinochet's Chile and Reagan's White House.

Friedman has a difficulty with which we can sympathize. At first, he is wary of discussing "assumptions" at all. Perhaps he learned this from studying Sir Isaac Newton, or from the solar installation expert on TV who said "beware assumptions." After seeing what von Thünen assumed under "isolation" we can appreciate Friedman's hostility.[47] Instead, Friedman prefers to discuss "implications" rather than "assumptions." Nevertheless, he says:

> ..the relevant question to ask about the "assumptions" of a theory is not whether they are descriptively "realistic," for they never are, but whether they are sufficiently good approximations for the purpose in hand.[48]

Having learned after von Thünen that assumptions are not about reality, Friedman is cautious about relating them to anything except "the purpose in hand." Since sooner or later this purpose is going to be related to reality, we see that he begins with an initial confusion. To be an "approximation" any **X** has to be a approximation *to something* and this something is going to turn out to be reality whether Friedman likes it or not. On the other hand, if his assumptions are other than relevant to reality, then they must be about something subjective—fictional or made-up. One can sympathize with any orthodox social scientist subjectivist who would like to be of this world and yet not be of it. It is well-expressed in the American popular song: "Make the World Go Away".

The theoretical tradition of subjectivism enjoins unreality; practical tradition, on the one hand, encourages, demands, reality. Friedman's training, such as it is and such as it was allowed to be under subjectivism, checks him from coming right out and saying baldly that what orthodox economics is all about is equations and descriptions of subjective states of the minds of non-economists (since they have no objective basis) posting as economists. Like John Dewey, who probably initiated this style of thinking at the University of Chicago having firmly established it earlier at the Universities of Michigan and Minnesota, Friedman quickly transforms "reality" into "experience." Experience may be *of* reality, *about* reality, but experience is *not* reality. When I say that I am *experiencing* reality, this is quite different from my saying that I am experiencing experience. If you believe that the sun was out there before human beings experienced it, then you believe in objectivity and the material world and you can be taught the principles of science and scientific method. If not, not. Friedman proves that he is an idealistic subjectivist by saying: "..the main issue—the conformity to experience of the implications of the marginal analysis is the issue as to whether an adequate theory of monopolistic and imperfect competition can be set down."[49] It cannot be set down by such orthodox economists as Friedman because they have no reality to compare it with, to base it on. What they have is subjective experience compared with subjective experience. But to continue.

We must admit, Friedman says, "..the impossibility of testing a theory by its assumptions.."[50] Since a theory is largely made up of its assumptions, one might hold this to be analytic or redundant. I suppose that most people familiar with the materialist philosophy of science would admit that we cannot test theories by assumptions. Also we cannot test the laws and postulates that are included in theories by assumptions. We test a theory in practice in reality to see whether the events it predicts occur in reality. Friedman admits to something similar to this, for he says: "..the assumptions are false..because the theory does not work."[51] On the contrary, the theory does not work because the assumptions are false[52]. But, in a manner of speaking, front-back, past-present, down-up, effect-cause, and similar peculiarities don't really cut any ice because Friedman doesn't have here any reality (hence a space and time frame) to check his theory against anyway. However it is possible to check his theory against reality in Chile where inflation is 23% and unemployment 25%. As a major architect of Pinochet economic theory, Friedman must face this reality. (Source: S.J. Solarz, "South American Democracy," *N.Y.T.* 16 February 1984)

At this point in his book Friedman is getting steamed-up to his subject for he next says that "Despite the apparent falsity of the 'assumptions' [he has assumptions in quotes to let us know that they may at any time become "implications"] of the hypothesis, it has great plausibility because of the conformity of its implications with observation."[53] Observation of *what*? If the hypothesis is based on *apparently* false assumptions, maybe they are really true assumptions? Still hostile to assumptions, he next says: "..we could state the equivalent hypothesis without any apparent assumptions, in the form of a list of rules for predicting.."[54] To be sure, in which case you don't need the hypothesis because you already have an implicit theory which is a synthesis of confirmed hypotheses! It is quite apparent that Friedman is not clear in his own mind as to the differences among theory, law, hypothesis, assumption, and postulate. Yet he is teaching future economists who will uninterpret the capitalist unsystem. With such teachers who needs charismatic socialists to undermine the non-system?

What precisely is Friedman's view of science? He tells us:

> The ultimate goal of a positive science [not to be mistaken for negative science] is the development of a "theory" or "hypothesis" [we hope that he is not using these interchangeably!] that yield valid and meaningful prediction about phenomena not yet observed [or perhaps observed but as yet unhypothesized about?] ....Viewed as a language theory has no substantive content [why should we view it as a language? Also, theory has no substantive content in the sense that it is a description of reality which has the content instead]; it is a set of tautologies [Friedman must have picked up this fallacy from some Chicago logical positivist from Vienna]. Its function is to serve as a filing system for organizing empirical material and facilitating our understanding of it....[55]

Science could also *explain* already observed phenomena. Why should science be viewed as a language? Why not view it as an activity having feeling and cognition about reality expressed in language? In what sense is a theory a set of tautologies? A theory ties together, in abstraction, a group of laws in such a way as to serve as a description of some segment of reality. Theory is not merely a "filing system for organizing empirical material." Theory is actively used to direct investigation; it is not a passive receptacle, a notion that goes back to Locke.[56] Friedman's mechanistic idealistic model leaves out the active role of scientific endeavor. He thinks of theory as being contemplative and passive.

In the same chapter, Friedman says: "A fundamental hypothesis [here he should have used "assumption" or postulate since without it the investigation cannot proceed] is that appearances are deceptive.."[57] Nothing could be more false. The fight for science, from the time of Thales, has been based on the truth-like nature of appearance. The materialists who won the day over the idealists accepted this as the *sine qua non* of understanding the world. Lucretius gives us the classical mechanical materialist defense of it in *De Rerum Natura.* It was Plato and other anti-scientific and anti-materialistic eliteists who thought that the disreputable, dirty, senses deceive us and that special idealistic forms had to be mentally constructed in a transcendental realm of non-existing, non-objects. We are not surprised to find that Friedman, like all orthodox thinkers, subscribes to some form of idealism. Yet, we expect no more consistency from him in this than we would from a symbolist poet.

When we turn to the science of economics, we find Friedman saying the following: "economics as a positive science[58] [what is a negative?] is a body of tentatively accepted generalizations about economic phenomena that can be used to predict the consequences of changes in circumstances."[59] Of course, it is hard enough to predict changes without also predicting the consequences of changes. Yet he does give us some idea of what a powerful tool orthodox economics has forged for understanding the future! Exalted above most other "positive" economists, says Friedman, was "[Alfred] Marshall [who] took the world as it is [as a matter of fact he took the world to be his oyster, as an ideal construct of the mind like any other good orthodox economist worth his apologetic salt]; he sought to construct an 'engine' to analyze it, not a photographic reproduction of it."[60] Engines move things rather than analyze them. Yet, why should an orthodox economist know this since he is about as close to the productive process as is a guru worshipping in the Monkey Temple in Varanasi during the rainy season. What Friedman considers the importance to be of his phrase: "photographic reproduction" we are forced to guess at. It is probably a right jab at the correspondence theory of truth which states that a proposition or sentence is true when it "corresponds" to or reflects reality. By "reality" in the last sentence is meant the ontologically material world which is independent of Friedman's consciousness and will continue to exist long after he has disappeared.

Insofar as evidence for or against any hypothesis is concerned [Friedman sometimes uses "hypothesis" when he means law and sometimes when he means theory] "..the weight of evidence for or against a [sic] hypothesis can never be assessed completely 'objectively.'"[61] In other words, it can never be assessed, according to him. For scientific, objective, assessment depends upon doing the assessment completely objectively, not partially objectively and partly subjectively. That is what science is all about. Furthermore, considering that Friedman has proved to us that he hasn't the foggiest notion of what "objectively" means, the following comments are germane : first, he cannot know enough to make this statement which nevertheless may strike most other subjectivists as having a brain-softening plausibility. I mean by this that since he has no objective world he cannot have any objective test; second, the history of science militates against his view. The subjective brown-spots in scientific laws and theories, after the hypotheses have been confirmed in practice, are gradually or suddenly replaced by objective assessments as in Einstein's correction of Newton. The hypotheses of the designers of the Tacoma Narrows Bridge which collapsed some years ago were assessed completely objectively when the bridge crashed into the canyon below. There was no added subjective doubt in the motorist who watched his auto slide off the bridge to be dashed to pieces. I also maintain that the hypotheses and postulates of Friedman which I reject in this paper are objectively rejected, not only on the basis of materialistic ontology and histodiamat epistemology but also upon the logical basis that he cannot sensibly hold them as congruent with his other beliefs.

Nevertheless, and despite all of his confusion and distortion, much of which appears to have been inherited from the Viennese schools, Friedman makes an honest and significant admission which is well worth considering. It is the following:

> The weakest and least satisfactory part of current [orthodox] economic theory seems to be in the field of monetary dynamics [we can see why, with such prominent teachers of "dynamics" looking upon theory as a "filing cabinet"] which is concerned with the process of adaptation of the economy *as a whole* to changes in conditions and so with short-period fluctuation in aggregate activity. In this field *we do not have a theory* that can appropriately be called "the" existing theory of monetary dynamics.[62]

This states that Friedman believes that orthodox economics has no macro-economic theory of capitalism as a whole. What "whole" implies is an economic theory covering the practices of international capitalism, and in addition, feudalistic hangovers and socialist economic activity. If orthodox economists did cover these areas, then they could be said to have an economic theory "of the economy as a whole." Friedman is temporarily worried about piecemealism and fragmentation just as C. Wright Mills and I.L. Horowitz have worried about piecemealism in sociology. But—but—this fragmentation can only be eliminated when it is recognized that there is an objective reality out there to serve as a regulator. Reality cannot be constructed in consciousness without any controlling non-subjective norm. Before we close our discussion of scientific method, I wish to briefly describe the method of histodiamat, contrasting it with orthodox subjectivist "scientific" method, such as the employed by Friedman.

HISTODIAMAT METHOD AND ORTHODOX SCIENTIFIC METHOD

Even if Friedman knew orthodox "scientific" method, which is to say that method employed by bourgeois social scientists, he would still not be on firm ground. The major reason for this is that ordinary garden variety orthodox philosophy of science method makes insufficient or improper reference to historical considerations. Not only must there be an independently existing material world not only a correspondence between our thought and this material world, but also there must be awareness that consciousness reflects a social world historically developed and developing. Both the material world and the social historical world, here separated only for analytic purposes, are modifiable by man. Social consciousness reflects social being. True. In addition, however, social consciousness and social being at any given period are reconstructing the material and social world. The reality that is reflected in human consciousness is in process of development. While the ancient Buddhists knew this, the later ones often turned to their own form of subjectivism.

Histodiamat claims that its correspondence view of truth enables some kind of *direct knowledge* of the material and social world. For both Locke and Kant, on the other hand, knowledge is *indirect* and uncertain. Kant, however, improved upon Locke in that Locke believed that the mind *passively* receives impressions and takes on an active role only in assembling simple images into more complex ones. What is missing in both Locke and Kant is the practice in which the contradiction and unity of development is brought out to indicate social development and revolutionary change.

In addition to correspondence-dialectical epistemology there are also formal truths that reflect general laws of thought and material development. One of these is that a thing cannot both be itself and not itself at the same instant. Dialectical logic, which complements formal logic, modifies this. For dialectics, a thing is not only what it is at time t, but is what it is becoming, what it is going to be. Dialectics added to logic what calculus added to mathematics. It made possible an analysis of the rate of change [of direction]. Another truth is that things develop by means of opposition. These are instances of formal logic and dialects taken together. The laws of science indicate the general characteristics and development of types of phenomena in reality. Such laws require a theoretical framework in a way that a particular established fact, say that Democritus the Greek materialist lived to be ninety-nine, does not. The laws and the fact are true but in different senses. So what we have are formal validity, laws of science (theory), and established truth. At any given time the truths that we know are partial in relation to absolute truth. Absolute truth is a limit concept which defines "the logical possibility of infinite progress in knowledge."[63] Absolute truth is heuristically important to make it clear that *relativism* is not viable ultimately.

Thought may go from the concrete to the abstract (or reverse), from the less comprehensive to the more comprehensive. Abstractions reflect nature more completely than concretions because they take into account more inter-connections.[64] Knowledge remains unscientific, but not necessarily false, if it is not related to a theoretical framework which makes possible the discovery of new implications. If this were not the case primitive man would have known nothing which is of course absurd. The student already knows truly a great many things before he is formally exposed to scientific theory. When I worked as an installation electrician at a shipyard on the North American West Coast, I was sent to school to learn the theory of marine electricity only after I had had months of practice in installing cable and hooking-up circuits. What I learned in the class on electrical theory was how these disparate operations that I "knew" were tied up in a general and abstract way. I caught myself saying: "Oh, I see, now. Well, that means that. ." and so forth.

We return now to the continuation of the attributes of subjectivism, focusing next on relativism to which I just referred above.

RELATIVISM

Relativism, which goes hand in hand with subjectivism and the refusal to admit a knowledge of an objective world, follows necessarily from the assumption that objective reality cannot be known and that what we know is only individual experience. It is the basis, as Crosser points out, of the German Historical School and American Institutionalism. As he says:

> Both schools are anti-conceptual in the sense that their very relativistic cognitive basis effects of necessity a fragmentation of thought. Both schools of thought emphasize empirical research and shy away from generalization.[65]

Relativists do not select data with regard to a specific socio-historical pattern but approach problems eclectically, now introducing anthropological factors and now psychological ones. Thorstein Veblen, for example, emphasizes change of attitude as a major factor in deciding intellectual affairs. Fustel de Coulanges in *The Ancient City* says: "Man has not, in our day, the way of thinking that he had twenty-five centuries ago; and this is why he is no longer governed as he was governed then." Yes—but why? C. Wright Mills reports that the empirical social scientists, the relativists, shift their emphasis from the history of institutions and ideas to the concrete behaviour of peoples. Studying one isolated sector of human affairs has been the growing tendency for sociologists and other social scientists. They place correlated emphasis upon studying what is contemporaneous rather than historical social events that cover a significant length of time. Such emphasis underscores relativism which looks upon each area or problem as unique—as without interconnections with other areas, the opposite of the integral dialectical approach. Too often resulting in the most shallow kind of relativistic explanation, this approach consists in collecting of data and the operation of statistics upon them without reference to theory, large social units, or history. Relativism is connected to the question of relevance. Trying to do significant work, the social scientist finds himself faced with the question of the relevance of his work to society and history. If he is aware that his kind of relevance is going to be rewarded by non-promotion or actual dismissal, he feels that it is safer to carry on relativistic research and commentary unrelated to any social goal except accumulating data. Relativists use as an excuse the fact that they are responsible not to society as a whole, but to the agents or administrators who pay them for doing research or teaching. Many nuclear scientists have said the same. As Mills puts it "..the economics of truth—the costs of research—seem to conflict with the politics of truth—the use of research to clarify significant issues.."[66] Relativism, it appears, is not only the result of subjective non-ontology, but also the pressures of bourgeois institutions.

FORMLESSNESS

If our knowledge depends upon individual experience, it is difficult to discover a pattern to reality. The lack of system among subjectivists is proverbial. Piecemealism is the accepted norm with area and problem remaining unintegrated into a large scheme of explanation. Purveyors of some form of subjectivism, however, have tried to overcome this atomic formlessness by constructing a molecular pseudo-reality. Differing from scientific mental structure in that it rejects the objective material world, subjectivism replaces the objective with a subjective ideal world. One example of this is the replacement of the material world by physicalism (which has nothing to do with the physical=material) in the constructionism of logical postivism or the replacement of material objects by unspecified "objects" of a mental order. At best this can lead to the construction of a fictional world built up in an apparently logically rigorous way. J.H. von Thünen is given credit for having begun the "isolated factor" approach which makes much of economic theory in the past hundred years read like a catalog of assumptions with little systematic connection with larger objectivity. "Let us assume that black is white"; "let us assume perfect competition"; "let us assume that all workers are more prosperous than over they were before"; "let us assume that capitalists now have a keen sense of social responsibility" and so on. Having assumed the "isolated state or factor," von Thünen draws out the implications that are assumed to follow in his make-believe world. If *Gulliver's Travels* had been generally assumed as an "isolated" basis for economics, then Swift could have drawn logical consequences and published them as economic science if he had not had certain intellectual and material standards based upon a respect for reality. The consequences of this method are to separate, isolate, economic thought from social and economic reality. As Crosser points out, the difficulties of this approach include that "If the inner recesses of the human mind are to be permitted to function as the sole guides of economic reasoning, there is no way left by which mental aberrations can be distinguished from normal mental processes, as Menger himself admits."[67] The formlessness that follows from not knowing reality is replaced by formalism in which everything is assumed even though it may have little or nothing to do with socio-economic reality. What the formalists have done is to develop a "rigorous" system with no realistic content. Much irrelevance and emptiness has been covered up by mathematical symbols and equations. It may sometimes take a decade for a "bright" neophyte studying the works of "isolationists" and "fictionalists" to discover that what he has been carefully learning is totally irrelevant to reality. Most of the budding economists, unfortunately, merely continue to assume that what they are doing has some mysterious descriptive value. Today in subjectivist philosophy, formalism consists largely of the reduction of what is normally thought of as philosophy to the logical analysis of language with the aid of symbols. The extreme subjectivism introduced with much trumpet-playing fifty years ago has been somewhat modified by some logical empiricists who now claim to need metaphysics—yet still without objective nature.

INDEPENDENCE AND PURITY

Much as in "isolation" the state of "independence" consists in denying that intensive and extensive knowledge of the socio-economic historical world is necessary to carry out significant theoretical work in economics. Independence separates inner forces from outer factors and subjective from objective "states". We have already seen that subjectivism views society, not as a whole, not as an historically conditioned organism, but as an agglomeration of individuals. Here is the initial separation and "isolation" that leads to the postulation carried on by von Thünen and his successors like David Gale. Franz Brentano the Viennese founder of much subjectivism thought that whereas outer experience is unprovable, inner experience is the basis of certainty (such as we can have it). Furthermore he held that psychology should be purely descriptive instead of causal and explanatory, thus separating ethics from other branches of social science. Just as they are held to be the source of our "knowledge" of the real but unknowable world, for Brentano sensation is the source of psychic life.

Even before Brentano, there was a shift in the significance of the word "real" until it was either put in quotation marks by orthodox social scientists so that they didn't feel compelled to explain what they meant by it, or it was assumed among social scientists to refer to a secondary reality. "Formal" reality was still primary for them.

When the Austrians and Germans particularly placed emphasis upon "purity" we find another touchstone of subjectivism unveiled. Von Wieser finds experience presenting data to economists as "a fountain head so inexhaustible in its riches, so dependable in its *purity*."[68] Purity is possible, pure theory, "because we maintain independence from time and place."[69] Husserl finds it indispensable to "retire to the realm of 'pure' thought."[70] This is equivalent to being in a state of "pure" immanence which finds its "place in transcendental subjectivity." It is first taken, Farber says, as a methodological expedient, but it soon becomes reified or "frozen" into a metaphysical non-natural domain. All philosophical problems for Husserl end up in the reduced realm of pure experience, according to Farber. We should not forget that Kant's most famous book is entitled *The Critique of Pure Reason.* Even pedestrian textbooks in elementary economics refer to "pure" competition and "pure" monopoly, "pure" investment, "pure" financing project which means that this concept has moved from theory into what are relatively mundane affairs. The ease with which "purity" moves from primary orthodox economic theory to secondary theory is quite astonishing and gives us one clue as to the impractical and unrealistic nature of orthodox economics even when it tries to get down to the vulgar nitty-gritty everyday tactics of micro prediction.

As we might expect, as we approach the Viennese heart of orthodox economic theory, we hear more about "purity". There is no reason to doubt, I think, that Husserl's inclination in that direction was suggested by Brentano. If we turn to Schumpeter's voluminous *History of Economic Analysis* (1957)[71] we find that his account of "General Economics from 1790-1870" is entitled "Pure Theory". As Schumpeter notes in this work, the tendency to abstract from the relations of production led, in post-Ricardian economics to the belief "that the economic process, however much given to stalling under the impact of 'disturbances,' was yet free, in its pure logic from inherent hitches."[72] Precisely what "pure logic" could mean in this context since such logic implies deduction within the limits of postulated formalism, is difficult to say except that it is a hortatory word giving the impression that theoretical independence and isolation are legitimate.[73]

The relationships between "purity" and "reality" need further examination but we note that even before Brentano there is a shift in the meaning of the word "real" until it is either put in quotation marks by many orthodox social scientists so that they don't have to explain what they mean when they use it, or it is assumed among them that when they use the word "real" they are referring to a secondary mental or financial reality. There seems to be not the slightest sense of modesty in arrogating to themselves powers given only to characters in fairy stories. Each one seems to have his own lamp to rub. In the meantime the real world of matter has been discarded as unknowable—and good riddance too. As was beautifully written in the "Admonitions of the Instructress to the Court Ladies" (Wei) "In nature there is nothing high which is not soon brought down"; there is nothing pure that is not soon impure. At the heart of the symbolist movement in France in the 1880's and 1890's is the literary critic-philosopher Remy de Gourmont. He said that "we no longer have principles, and there are no longer models; a writer creates his esthetic in creating his work.."

We now move from metaphysics to sensuality in which general principles are reduced to personal impressions and pleasure. One will be "able to erect personal impressions into law." All this will be "complex in origin and pure in quality." In reflecting on the meanings of "purity" and its connection with *katharsios* in Greek and *purus* in Latin, it became apparent to me that the classical meanings lingered on as sort of a cobweb on the flower. But more particularly (the meaning of *idiot* (idiotic=private) also comes to mind), the meaning of "pure" in the nineteenth century history of subjectivism is to be specialized specialization. "He is pure lawyer, pure doctor, pure dancer, pure poet." What this means is that he is so completely specialized that he approaches idiocy=complete privacy in his monomania. Marx makes references in *Capital* to the specialization that by necessity accompanies capitalism. The necessity is to reduce all operations of production to the simplest (hence most specialized, tasks). Pure economics, then, is the most specialized economics and hence also partakes of the idiocy of specialization. When we say that he knows more and more about less and less it tells us that he is becoming more specialized, more privatized, and less socialized and less objective. Much time and energy of young would-be social scientists goes into learning this double-entry ontology. With such an outlook do they have enough time or incentive to discover reality ? I cannot see that from the point of view of Occam's Razor there is any difference between this and the medieval disquisitions on the question as to how many angels can rest on the head of a pin which has the virtue of having ontological implications. Today it could be raised as to how many fictions can rest in the head of an orthodox economist. This lamentable state of affairs accounts in part for the unproductive, non-predictive work of social scientists whose output becomes, at best, formalistic and isolated, idealistic apologetics for the status quo or at worst a resuccitation of unworkable rejected theories such as proposed by Milton Friedman to be used in Pinochet's Chile and Reagan's U.S.A.

TIMELESSNESS AND NEUTRALITY

Timelessness occurs when it is believed that theory need not have concrete ontological moorings in a given society having a time, place, and history. Marvin Farber's criticism of Husserl's *Lebenswelt* is that it cannot have transcendental existence simply because in order to have any kind of existence it must have a time, place, and *be something*. As old as Parmenides, timelessness has the same value today that it had in his day. Essentially, timelessness is a rejection of history, which is tantamount to the rejection of the external world as knowable reality. Ernest Gellner in *Thought and Change* finds the first instance of modern timelessness in the distorted Hegelianism of F.H. Bradley.[74] If we find an academic avoidance of social implications, we also find timelessness. Associated with neutrality, it is also to be found in logical positivism and logical empiricism.

Neutrality is another substitute for the objectivity which has vanished with the external material world. Beyond the mundane, neutrality is predicated on timelessness when it is found necessary to postulate formal schemes and equations which are to hold good for *any* [therefore *no*] society.

Michael Polanyi, according to Joseph Needham, "seems to take the view that the laws of 'fortuitious concourse of particles' have the status of laws eternally applicable to human society."[75] Timelessness occurs in anarchic "social" atomism. For Polanyi, order rises spontaneously from the interaction of the particle-laden dynamic field. The analog of this, Polanyi finds, is the competitive atomistic capitalist order. He also oppossed scientific planning in the name of "freedom."

> ..the fundamental drawback of atomistic capitalism is that its parts have no organic connection with the whole. The only concern of the industrial monads, is not that cost and price should balance, but that the latter should outweigh the former by as much as they can make it. Hence a condition of universal war. [a la Hobbes, one of the first to note it].[76]

Finding words of the poet Auden to the point here, Needham quotes him with reference to the bourgeois centuries as follows :

On sterile acres governed by
Wage's abstract prudent tie
the hard self-conscious particles
Collide, divide, like numerals
In knock-down drag-out laissez faire
And build no order anywhere.[77]

As Hobbes [who lived to be ninety-one] said, universalizing from the capitalist unsystem [the poor] man's life is short, brutish, and nasty. About neutrality, Robert Briffault pointed out somewhere in *Reasons for Anger* that one is either "neutrally" for something or "neutrally" against it. In the opinion of U.S. Supreme Court Justice Louis Brandeis, "Neutrality is at times a graver sin than belligerence."

FROM TIMELESSNESS TO THE INSTANTANEOUS

Philosophy, as opposed to science, correctly begins, according to Henri Bergson, with the intuitions of the inmost reality of consciousness which is experienced as duration. Duration becomes the cornerstone of all that makes philosophy different from, and superior to, science. Scientific knowledge, according to Bergson, has its origin in the vulgar knowledge of ignorant people and children. True or sophisticated knowledge consists in the inner unity and integrity of the soul which approaches purity rather than vulgarity. What this means, according to Paul Nizan, is that :

> ..despite the supercilious air of detachment and remoteness [of such bourgeois intellectuals as Bergson]....are all totally obsessed with that passive but very real pleasure which a bourgeois experiences when he contemplates his own person.[78]

What Nizan says is fittingly ascribed also to the American school of subjectivist philosophy called Personalism, not to mention Existentialism and Phenomenology. As the Father of American Personalism said: "Finally, we emphasize the futility of all attempts at philosophizing on the plane of impersonal existence. On that plane thought is blocked in every direction. If we seek for explanation we never find it. Things themselves are dissolved away into elusive phantoms."[79] What is said here is that we cannot know the external world, and therefore must become reconciled to settle down as best we can into the virtual solipsism of contemplating our own marvelous, albeit tortured, self within the specious flux of appearance. Since no one describes the bourgeois intellectual better than Paul Nizan, let me add one more passage from his ingenious and educative book :

> The bourgeois is a solitary individual. His universe is a world of abstract mechanisms, of economic, juridical, and moral relationships. He *does not come into contact with concrete reality*; he does not interact directly with other men. His property, too, is an abstraction. He is far removed from events. He sits in his office or his bedroom surrounded by the little coterie of objects which he is so dependent upon: his books, his papers, his desk, his bed, his wife. . . .He is thoroughly insulated from the natural shocks of the world[80]. His culture consists entirely of screens and shock-absorbers; it is a dense network of intellectual schemes, an exchange of signs and signals. . The bourgeois believes in the power of titles and words; he believes that anything called into existence will be, provided only that it be given a name.[81]

Duration is truer, deeper, and shorter, than scientific time (knowledge) according to Bergson. This is perfectly in harmony with the view commonly attributed to the mystical philosophers of India. In the next generation, after the first World War, Bergsonian subjectivism reappeared in Heidegger and then in Sartre. Even so, it was already present in French philosophy as early as 1876 in Amiel. He had already enunciated a cardinal doctrine of subjective existentialism when he wrote : "The absurd is the very nature of life. . . . ."[82] Of course life is absurd if we can never know reality!

Under objective, scientific, claims, that is to say, materialist claims, a thing (good, utility) may have a value at any given time as long as it embodies labor power. This takes care of the common sense, vulgar, temporal, dimension of reality that has bothered elitists from Plato to Heidegger. On the subjectivist view, things and values pop in and out of existence depending upon whether they are subjectively desired under the aspect of duration, or inner time. This belief accounts for much of U.S. foreign policy, which is famous for being guided by impulse.

PSYCHOLOGISM

As C. Wright Mills defines it, psychologism "refers to the attempt to explain social phenomena in terms of facts and theories about the make-up of individuals."[83] Mills' claim that this "rests upon metaphysical denial of the reality of social structure,"[84] should be rephrased to read "rests on the assumption that there is no objective structure to social reality" because "metaphysics" could be consistent with either view. If the word "objective" is left out it will be possible for Mills to fall into some fictionalist, formalist or, pragmatist trap. As Herbert Aptheker has pointed out:

> Mills seems to believe that scientific methodology and scientific laws are the result of purely pragmatic experiments, and that the laws and methodology are not fundamental determinants of the experimentation. Thus, he believes "the epistemology of science is parasitical upon the methods that scientists, theoretical and experimental, come to use.[85]

We have already mentioned the psychological "substructure" hypostatized by the orthodox economists belonging to the "First Austrian School." Much orthodox economics changed from an objective science, under the classicists and Marx and Engels, to a branch of subjectivist psychology. Emphasizing the need to understand the consumer and his attitudes becomes a major branch of economic psychology or psychological economics, even though von Mises points out that their relationship is still fluid. Psychologism, since it has no external objective world to examine, must be content with information taken from samples of individual volition. Mills sees one cause of psychologism to be the "lack of firm connection with substantive problems."[86] Mills is correct. It is hard to have substantive problems without any substance.

Recently I had the pleasure of hearing a television sports announcer exclaim after a brilliant shot from a sandtrap to the green had been executed by a renowned professional golfer: "It's not a matter of backswing, it's a matter of character !" Why stop there ? Why not extend this kind of psychologism to the universe in which it is not a matter of material particles in motion but the psychological set of the man upstairs, the Great Golfer or supermentality? Salesmanship, which has replaced much social science, leads to a desire to be able to know what the customer or consumer wants so that we can sell him something. "Science" then becomes a method by which to manipulate the consumer.

IDEOLOGY OF THE FIN DE SIECLE

After spending a great deal of precious time reading the

disconnected statements of orthodox economists [what they say and write cannot be called theory because they do not have the base or coherence required to have theory] I have come to the conclusion that there is no orthodox economic ideology as such—at least not as an articulated body of beliefs. Ideology, like theory implies coherence, systematization, and some substantive grounding. What one finds in so-called economic theory is not an ideological system, but subjectivist confusion. Who needs bourgeois ideology when he already bows before bourgeois confusion ?

The root cause of the confusion is the inability of orthodox thinkers to solve the basic contradiction of capitalism. This contradiction is that capitalism has increased the productive power of man to the point where many of his needs can be taken care of. Yet at thee same time, the unsystem requires, because of is peculiar development, the expansion of profit. The system has burst its seams in one direction and shriveled up in another. Uncertainty reigns in this anarchy. Productivity and salesmanship are separated from ethical considerations. Planning for a future improved system continues to be regarded as pernicious. Ontology and ethics are replaced by fictions. Philosophy becomes useless except to add to the confusion in a pompous manner and then spread its own disarray and jumble. Under a relatively rational and realistic condition, philosophy can be a science of changing reality. Reality is something that can be objectively known. Today most philosophy seems to be the non-science of what cannot be objectively known. The confidence and hope placed in logical positivism and empiricism withered up. What was going to be the UNITY OF SCIENCE turned out to be the UNITY OF SUBJECTIVISM. The symbolic capers and trivial mathematization of formalism replaced the description and assessment of reality.

SUMMARY AND CONCLUSION

Modern subjectivism finds its economic stimulus in providing an apology for eroding capitalism. Its main superstructural model was provided by the subjectivism of Kant. Kant, in order to save God, immortality, and freedom, attacked the material foundation of science which seemed to be responsible for eliminating them. Neo-Kantianism, an almost all-pervasive subjectivist philosophy, wears a double mask. One side of it is fideist, although there are progressive elements in Kant.[87] The other side is anti-materialistic, anti-objective and by the end of the nineteenth century, also anti-socialist. Scientific socialism requires the postulates of materialism and objectivity which are denied by subjective idealist forms of pseudo-science.[88]

Economics was the first social science to become institutionalized. It developed in two directions: the first, socialistic, and the second anti-socialistic, or orthodox. Practitioners of orthodox economics, in return for avoiding a materialist (objective) description of reality were rewarded by steady employment, research funds, increasing applause, and Nobel Prizes. The slightest substantive criticism of capitalist economics (as by J.K. Galbraith) has led to prompt rejection in Sweden. The apologetic, subjective stance of orthodox social science in general has not gone unnoticed among the more critical, however. The literature of protest and criticism has perhaps doubled every decade since 1890.

The major characteristics of subjectivism in social science have been idealism, fideism,[89] anti-materialism, anti-science, anti-socialism, anti-objectivity (avoidance of ontology), anti-historicalism, anti-systematicism (unless system is formally unbased on non-ontology) replacement of "truth" by validity, isolationism, neutralism, and anti-intellectualism.[90]

Subjectivism is being more openly attacked today because its intellectual and financial support is weakening. This has come about because the contradictions of capitalism are becoming so obvious that only the blind can avoid seeing them. The relatively great homeostasis of socialism since the second World War has also been difficult to hide.* Five out of six bloody conflicts are capitalist-subjectivist in origin. Histodiamat, in the meantime, poses a systematic and continuous critique of subjectivism and exposes its fraudulent and opportunistic foundations and aims. Although past history suggests that subjectivism can continue in some form almost indefinitely, associated with reactionary institutions difficult to dislodge, its obvious inferiority to histodiamat is being more clearly seen just as the intellectual system of feudalism has lost its respectability and dignity in both social and intellectual life with the

*Despite the rabid hooper of the debased north America media.

advance of capitalism and socialism.

CONCLUSION

Our survey of both Indian and Western philosophy of science has clearly revealed the subjectivism inherent in it. Subjectivism at first was moored largely in a harbor of ignorance. But its increased expansion became evident in the eighteenth century in the struggle between the Enlightenment rationalism and scientism on the one hand and feudal vestiges, irrationalism and anti-scientific postures on the other hand. Despite the laudable work of Kant in historical astronomy and dialectics near the end of the eighteenth century, his attack on the ramifications of the scientific outlook in religion, epistemology and ethics led to his being designated, sometimes without justification, the leading philosopher of Western reaction. The question as to the source of the necessity that all logicians (including Kant) find in reasoning and logic (mathematics) is not answered by anyone who denies that we cannot know the independently existing material world. The necessity that is lost in subjectivism and which gives to subjectivistic arguments, accounts, analyses, their specious air is that very necessity which is based upon the operations of the material world independent of being experienced. Kant before his critical period produced salutary views; after his critical period he began to reason as an agnostic and subjectivist where the cognitive situation took precedence over nature as an object of analysis and speculation, and where constant whetting of the knife (Schopenhauer's phrase) seldom led to cutting anything.

No one is maintaining that Kant is directly responsible for mystical physics and mathematics or for "pure" subjectivist economics. His calamitous influence would have crumbled, however, in a rational society not overly determined by individualism and subjective aims such as profit-taking. His own contradictory views are considered to be the result of the relative backwardness of the German bourgeoisie of his day.

The Neo-Kantians looked upon scientific concepts and philosophical categories as logical constructions (Rickert) and decided that there was an unbridgeable gap between physical and social science (theoretical vs. practical reason in Kant), that social phenomena are insusceptible of scientific cognition. With relation to scientific socialism (histodiamat), neo-Kantianism through Bernsteinism* became the official doctrine of agnosticism and opportunism.

Today, the inroads of dialectical thinking have created a veritable clash of viewpoints between the highly educated know-nothing subjectivists and validitudinarians (those who believe it is sufficient to have systemic validity without truth). Social life cannot be based rationally on ignorance, indeterminism, agnosticism and formalistic gamesmanship. Philosophers of science must come to grips with the problems of objectivity and truth. We see many healthy signs in this direction although too many of them have appeared only East of Munich and south of Helsinki. We believe that the strength of American neo-Kantianism is subsiding and fresh new waves are bringing in long-neglected considerations of truth and objectivity.

---

*"The end is nothing movement is every thing".

NOTES FOR CHAPTER I

1 *The Aryabhatiya with the Commentary Bhatadipika of Paramadicvara*, ed. H. Kern (Leiden: Brill, 1874). See also George Sarton, *Introduction to the History of Science* (Washington: Wilkins and Wilkins, 1927-1948), I, 409.

2 See Marshall Clagett, *Greek Science in Antiquity* (New York: Abelard-Schuman, 1955), p. 90; Joseph Needham, *Science and Civilization in China*, Vol. 3: "Mathematics and the Science of the Heavens and the Earth" (Cambridge: University Press, 1959). For a condensation of Needham's view on Chinese astronomy, see his "Astronomy in Ancient and Medieval China", *The Place of Astronomy in the Ancient World*, ed. D.G. Kendall et al. (London: Oxford, 1974).

3 See V.S. Agrawala, genl. ed., *Vedic Mathematics* (Varanasi: Banaras Hindu University, 1965); S.N. Sen, A.K. Bag and S. Rajeswar Sarma, *A Bibliography of Sanskrit Works of Astronomy and Mathematics* (New Delhi: National Institute of Sciences of India, 1966). Sen dates Aryabhatiya at 449 A.D. See p. 7.

4 See Debiprasad Chattopadhyaya, *Science and Society in Ancient India* (Calcutta Research India Publications, 1977), p. 19f. "Speculations in the Medical Schools", Surendranath Dasgupta, *A History of Indian Philosophy* (Cambridge: University Press, 1932), II, 273-436 is still a mine of suggestion. Sarton in *op. cit.* mentions Caraka in I, 76f.

5 See Sarva Daman Singh, *Ancient Indian Warfare with Special Reference to the Vedic Period* (Leiden: Brill, 1965,) p. 23, 119.

6 See Radha Kumud Mookerji, *Indian Shipping* (Bombay: Orient Longmans, 1957), pp. 13-15.

7 *Ibid.*, p. 19.

8 (Calcutta: Indian Chemical Society, 1956). To be studied alongside this pioneering work is Ray's masterpiece, *Autobiography of a Bengali Chemist* (Calcutta: Orient, 1958). For guidance on the social dimension of scientific development, see A. Rahman, "Transformation of Scientific Movement to Social Movement," *Triveni*: *Science, Democracy and Socialism* (Simla: Indian Institute of Advanced Study, 1977).

9 See Abdur Rahman, "Lecture IV: Impact of British Science on India," (New Delhi: Research Survey and Planning Organization, 1968).

10 See Dale Riepe, "Selected Chronology of Recent Japanese Philosophy (1868-1963)," *Philosophy East and West*, Vol. XV, Nos. 3-4 (July-October, 1965). See also William W. Lockwood, *The Economic Development of Japan* (Princeton University Press, 1954), and Angus Maddison, *Economic Growth in Japan and the USSR* (New York: Norton, 1969). Also see Rahman, *op. cit.*, p. 18.

11 *Ibid.*, p. 19.

12 Compare this with the situation in the United States at American Independence. Many of the settlers in New England were from economically and technologically advanced Elizabethan England and "..[t]he motherland recognized the scientific merits of the gentlemen of New England, we can learn from the records of the Royal Society...."

Dirk Struik, *Yankee Science in the Making* (Boston: Little, Brown, 1948), p. 23.

13 In *Philosophy of Science: The Historical Background* (New York: Free Press, 1968).

14 *Conceptual Foundations of Scientific Thought : An Introduction to the Philosophy of Science* (New York: Macmillan, 1968).

15 See his *The Positive Science of the Ancient Hindus* (London: Longmans, Green, 1915).

16 *History of Chemistry in Ancient and Medieval India* (Calcutta: Indian Chemical Society, 1956).

17 Interview with Professor R.S. Gill, New Delhi (April 6, 1967). See his "Recent Development in the Philosophy of Mathematics," *Quest for Truth* (Delhi, 1976), ed. K.K. Mittal.

18 *Ibid.*

19 Rahman, Sen, and Rajagopal, "State Support to Scientific Research in India—An Analysis of Trends" (New Delhi: CSIR, 1966), p. 9.

20 Britain spent 80 percent of its scientific budget for military ends and the U.S.A. 90 percent in 1935. This has risen in 1981-85 under President Reagan. See J.D. Bernal, *Science in History*, p. 896. Bernal praised India for resisting the tendencies of scientific concentration practiced in Britain and the U.S.A. *Ibid.*, p. 898.

21 *Science in History*, p. 898.

22 Rahman et al., "State Support. .", *op. cit.*, p. 10. Other relevant papers of Rahman and colleagues include the following: Abdul Rahman, "Lecture IV: Impact of British Science on India" (New Delhi: Research Survey and Planning Organization, 1968). See also other works of Rahman such as:

Anatomy of Science (Delhi: National Publishing House, 1972).

with T.H. Shama Rau, *Flight of Scientific and Technical Personnel* (New Delhi: Research Survey and Planning C.S.I.R. 1966).

R and D Planning in the Framework of National Planning (New Delhi: Centre for the Study of Science, Technology and Development, 1978).

with N. Sen and K.N. Sen Gupta, *Scientific Journals in India: A Stude of their Characteristics* (New Delhi:C.S.I.R., 1965).

with K.D. Sharma, *Science Policy Studies* (Bombay: Somaiya, 1974).

Science Policy Studies in India: A Status Report (New Delhi, Center for the Study of Science, Technology and Development, C.S.I.R.; New Delhi: Council of Social Science Research, 1977).

Science Technology and Economic Development (Delhi: National Publishing House, 1974).

Survey and Planning of Scienctific Research Unit and Scientific Societies in India (New Delhi: C.S.I.R., 1965).

et al., *A Study of Expenditure in National Laboratories* (New Delhi: C.S.I.R., 1964).

Triveni: Science, Democracy, and Socialism (Simla: Indian Institute of Advanced Study, 1977).

23 Rahman, Sen, Rajgopal, *op. cit.*, p. 9.

24 *Ibid.*

NOTES FOR CHAPTER II

1 *Letters of Sri Aurobindo* (Second Series) (Bombay: Aurobindo Circle, 1939), p. 572.
2 *Ibid.*, pp. 572-3.
3 *Materialism* (Calcutta: Renaissance, 1951), p. 25.
4 See *A History of Warfare of Science with Theology*, 2 vols. (New York: Appleton, 1910), to refute Sri Aurobindo's contentions.
5 *Letters*, op. cit., pp. 571-2.
6 *Ibid.*, pp. 567. Post Mao Chinese nationalists are trying to reverse this.
7 *Ibid.*, pp. 563-4.
8 *Ibid.*
9 *Letters* (First Series) (Bombay: Aurobindo Circle, 1947), pp. 399-400.
10 Rosenberg, philosopher of the German Nazi Party, like Aurobindo, points out the failure of materialism as opposed to the correctness of the views of Kant and Schopenhauer because the latter two hold "to an undogmatic" idealism. See *Der Mythus*, p. 324f.
11 The British Crown appears to have a penchant for knighting idealists: Jeans, Ayer, Berlin, Popper, etc.
12 *Letters* (Second Series), p. 400.
13 *Ibid.*, p. 402.
14 *Ibid.*, (First Series), pp. 402-3.
15 *Ibid.*, p. 403.
16 *Ibid.*,
17 *Ibid.*, p. 367.
18 *Ibid.*, p. 254, p., 156. This notion hails back to the Earth Mother, to Matriarchal Times. See Robert Briffault, *The Mothers* (abridged) (London: Allen & Unwin, 1959), who says that"..the conception of Mother Earth is far more obvious than that of Mother-Moon. Even far above the level of primitive culture Mother-Earth plays no part. Both in Vedic literature and Hindu religion we find no great Earth-Mother, which may be caused by the fact that the Hindu Aryans developed their high civilization without even becoming agriculturalists. They established themselves in Northern India as warriors leaving agriculture to the native inhabitants, and despising it." p. 363. But Aurobindo could easily have picked her out of Bengali Tantra. The combination of material craftiness and other-worldliness that partially characterizes Sri Aurobindo is commonly found among the great mystics and psychics such as Meister Eckhart, Plotinus, Gandhi and Mary Baker Eddy. This combination of qualities may also inhere in the succession of primitive magician, priest-pharaoh, priest-aristocrat, priest-merchant and priest-computor magician.
19 *Letters* (First Series), p. 144.
20 *Ibid.*, p. 163.
21 *Ibid.*. p. 164.
22 *Ibid.*, p. 166.
23 *Ibid.*, p. 167
24 *Letters* (Second Series, p. 412. The ancient Taoists agree with this. See "Science, Magic, and Mystique of the Body," *Chinese Erotic Art*, ed.

Michael Beurdeley (Rutland: Tuttle, 1969). The origin of the belief that Nature or God engages in play may have had its origin from the dialogue between Badarayana and Kumarila. Kumarila said that God could not have created the universe because it has no purpose. Badarayana replies that it purpose is sheer fun and games. See Debiprasad Chattopadhyaya, *Indian Atheism* (Calcutta: Manisha, 1969), p. 224f. Nietzsche had a similar interpretation.

25 *Letters* (Second Series), p. 414.

26 These references to sex were written by Aurobindo when he was approaching his mid-sixties. It is doubtful whether he discussed this with a yogi. This lapse may have been caused by his being unable to find one. Dom Denys Rutledge in his *In Search of a Yogi* (London: Rutledge & Kegan Paul, 1962), claims to have searched throughout India for a yogi who had supernatural powers, but found only yogis often bereft of natural ones.

27 It is gratifying to learn that Aurobindo, in his low opinion of science, consistently fails even to interview one yogi on this interesting problem.

28 *Letters* (Second Series), p. 418. Much depends on the muscles of the *dobhi*, from the Sanskrit *dhona* "to wash."

29 *Ibid.*, pp. 419-20. Taoist magicians take note !

30 *Ibid.*, p. 420.

31 See *Cutting Through Spiritual Materialism* (Berkeley: Shambhala, 1973), p. 25.

32 (Calcutta: Progressive, 1955).

33 *Ibid.*, p. iii.

34 *Ibid.*

35 *Ibid.*

36 *Ibid.*, p. 3.

37 *Ibid.*, pp. 7-8.

38 *Ibid.*, p. 14.

39 Lewis Carroll, "Alice's Adventures in Wonderland," Chapt. VIII in *Logical Nonsense: The Works of Lewis Carroll* (New York: Putnam's Sons, 1934), p. 122. According to the Sixth Patriarch, "Hui Neng has no means to cut off any thoughts," R.H. Blyth, *Oriental Humour* (Tokyo: Hokuseido, 1963), p. 93. This is because he cannot cut off the external world which is the source of sensations and abstractions.

40 *Philosophy of Science*, p. 172.

41 *Ibid.*, p. 173.

42 *Ibid.*, p. 172.

43 *Ibid.*, p. 173.

44 *Ibid.*, p. 181.

45 *Ibid.*, pp. 181-2.

46 (Bombay: Asia, 1962).

47 *Ibid.*, p. x.

48 *Ibid.*

49 *Ibid.*, p. 2.

50 *Ibid.*, p. 3.

51 *Ibid.*, p. 4.

52 *Ibid.*

53 *Ibid.*

54 *Ibid.*, p. 5.

55 *Ibid.*

56 *Ibid.*

57 Isaac Newton, *The Mathematical Principle of Natural Philosophy* (New York: Daniel Adee, 1846), p. 1.

58 *Philosophy and the Physicists* (London: Metheun, 1937), p. 266. After having noted this, Stebbing takes a neutralist position between idealism and materialism, unconscious of the implication of doing it—namely that she herself falls into subjectivism and idealism because she has no ontology. In this she followed the lead of Philipp Frank. Frank had no use for idealists who claimed that modern physics gave new evidence for spiritualism or idealism. His tendency towards subjectivism may be seen in his statement: "Everything in the world that is describable through intersubjectivism expressions is called "matter" by diamat [dialectical materialism]." This is of course without ontological basis and hence could mean that idealistic and totally subjectivist utterances could be called "matter." Frank, *Modern Science and Its Philosophy* (Cambridge: Harvard, 1949), p. 201. Although Frank is more sympathetic to histodiamat than Stebbing, who presumably knows very little about it, his interpretations of it turn it into a subjectivist philosophy.

59 See *Albert Einstein*: *Philosopher-Scientist*, ed. P.A. Schlipp (Evanston: Library of Living Philosophers, 1949), p. 470. This work should be read *ad libidum* because it is filled with humorous *apercus* provided by many idealistic physicists.

60 *Philosophy of Matter*, p. 231.

61 *Ibid.* Let us be warned that when philosophers begin to speak of the "finite" that they are only opening the door to speak of the "infinite" which they often regard as a *place*.

62 (Santiniketan: Centre of Advanced Study in Philosophy, 1965).

63 *Ibid.*, p. 150.

64 Quoted, *ibid.*, p. 4.

65 *Problem of Fact*, p. 151.

66. Pandey, like Daya Krishna and other Indian philosophers under the spell of British linguistic philosophy, has evolved into a state of awareness of quaness.

67 *Problem of Fact*, p. 153.

68 *Ibid.*

69 *Ibid.*, p. 154.

70 Like focusing attention on the unforeseeable

71 Leo C. Rosten defines "freedom" as "to be without money." "Infinite freedom" would be an infinite extension of that.

72 *Problem of Fact*, p. 156.

73 R.G. Collingwood in *Faith and Reason* makes considerable sense until he introduces this unfortunate term with all its theological implications.

74 *Problems of Fact*, p. 156.

75 Idealists like to deal indiscreetly with the "undifferentiated," because, since it has no known qualities, either nothing or everything may be said about it. F.S.C. Northrop, the American comparative philosopher makes the "undifferentiated" one of the two basic elements in his study entitled *The Meeting of East and West*, (1946).

76 *Problem of Fact*, p. 156.

77 (Kailasapuram (n.p., 1966).

78 *Concept of Matter*, p. iii.

79 Not the celebrated South Indian dancer.

80 *Concept of Matter*, p. 11.

81 *Ibid.*, p. 43.

82 The philosophers who have disappeared into this bog are barely denumerable. See Kao Hsing-hua et al. In "There is also Identity between Erroneous Thinking and Existence," from *Kuangming Daily* (August 5, 1960), *Chinese Studies in Philosophy*, Vol. III, No. 4 (Summer 1972).

83 But Gauss was able to represent complex numbers by points on a plane to indicate a kind of non-imaginariness. See Struik, *Concise History of Mathematics*, p. 142. Fantastic things are combinations of things found in reality.

84 If Freud is correct, the same is true of dreams. In addition to this, according to Freud.". . no dream is inspired by other than egoistical motives." *Interpretation of Dreams* in *The Basic Writings of Sigmund Freud* (New York: Modern Library, 1938), p. 420. The relationship between fantasy and magic requires more research.

85 "Being and Structure," tr. from the German by John C. Cullen, *Revolutionary World*, Vol. III (1974), pp. 46-7. If Varadachari had been a Spinozist we could perhaps more readily accept his view of unification.

86 "Two Schools of Materialism in Indian Thought," *Vidyodaya J. Arts, Science Letters*, Vol. 2, No. 2 (July 1969), p. 87.

87 Tr. by Kenneth K. Inada, *Nagarjuna* (Tokyo: Hokuseido, 1970), p. 84.

88 "Influences of Industrialization and Technology on the Philosophies of India," *Prabuddha Bharata*, Vol. 62, Nos. 7-8, p. 253.

89 Tiwari, *Secularism and Materialism in Modern India* (Delhi: Banarsidass, 1964), p. 319.

90 *Ibid.*

91 *Ibid.* For Lenin's view see *Materialism and Empirio-Criticism* (New York: International, 1970), p. 294.

92 3rd ed. (Princeton: Van Nostrand, 1958), p. 1035.

93 Ed. E.B. Uvarov & D.R. Chapman, 4th ed. rev. Alan Isaacs (New York: Schocken, 1972) p. 237, pp. 132-3. These careless definitions give us another clue as to why there is so much confusion concerning the meaning of matter. The first one, by Van Nostrand, is circular as well as hedging. If one is going to make a distinction such as "From the classical microscopic point of view," he ought also to give a few alternatives, including some macroscopic view. Respecting the definition of energy, Penguin makes it appear that energy can sometimes be apart from matter—what they have in mind is radiant energy. But

how can there be energy which can do work apart from matter? This is confusing and unscholarly. It surely indicates the banal influence of idealistic science education. The definition of matter by C.D. Broad of the University of Bristol may be as complete as any. According to him matter is independent of minds, observed by many minds, neutral with respect to the senses, having sensible qualities, has considerable persistence, has position in space and is capable of motion. See his *Scientific Thought* (New York, 1923), pp. 229-234.

94 *Readings in the Philosophy of Science* (New York: Appleton-Century-Crofts, 1953), p. 382.

95 *Ibid.*, p. 384.

96 *Philosophical Papers* (New York: Colliers, 1962). See "Proof of an External World". *Philosophical Studies* (New York: Harcourt, Brace, 1922), pp. 1-3.

97 *Ibid.*, p. 30.

98 Moore hints at this when he says that some people might want a "stronger argument than the one about the hands mentioned above. This would require "premisses which are not known to be true, unless we do know of the existence of things", "Proof of an External World," *op. cit.*, p. 148.

99 "Refutation of Idealism," p. 30.

100 *Ibid.*

101 Tiwari, *op. cit.*, p. 319.

102 *Ibid.*

103 According to Lenin in *Philosophical Notebooks*, tr. Clemens Dutt (London: Lawrence & Wishart, n.d.): "Materialism is the son of Great Britain. Its scholastic Duns Scotus had already raised the question '*ob die Materie nicht denken konnte*3' [whether matter couldn't think]. He was a nominalist. Nominalism is in general the first expression of modern materialism," p. 321.

104 Engels, *Dialectics of Nature* (London: Lawrence & Wishart, 1946), p. 320.

105 *Secularism and Materialism*, p. 319.

106 *Ibid.*, p. 323.

107 Lenin's citation in his *Materialism, op. cit.*, p. 296.

108 Tiwari, *op. cit.*, p. 325.

109 Lenin, *Materialism*, p. 140. That is, a vital leap.

110 *Op. cit.*, p. 342.

111 *Ibid.*, p. 342-3.

112 *Ibid.*, p. 343.

113 Putnam, *Mind, Language and Reality* (Cambridge: University Press, 1975), II, 210.

114 Tiwari, *op. cit.*, p. 339.

115 *Ibid.*, p. 346. Tiwari notes that "In epistemology, Roy adopts some of the doctrines of the modern American materialists, Sellars and others, *and* of the Logical Empiricists." [My italics—DR], p. 350. This gives one a clue to the fertile confusions fostered by M.N. Roy, for Sellars was a materialist while the logical positivists were subjective idealists.

116 (Calcutta: Academic Publishers, 1968).

117 *Ibid.*, p. 5.
118 *Ibid.*, p. 10.
119 *Ibid.*
120 *Ibid.*, p. 149.
121 *Ibid.*, p. 148. In the opinion of Daniel Goldstick, "Popperians can only conceive of the history of science as a trail littered with the whitened bones of falsified theories giving us no reason to suppose current scientific theories will not in their turn encounter like refutation and hence giving us no reason to accord any practical *belief* to what these scientific theories say about reality." "Reading Althusser", draft of unpublished paper (Toronto, 1973), pp. 3-4.
122 Sengupta, *op. cit.*, p. 149.
123 *Demonstration and Logical Truth*, p. 179;
124 *Ibid.*, p. 200.
125 Hilary Putnam, *op. cit.*, II, 213. Carnap like Sengupta never gets out of the formal squirrel cage. Putnam argues that". . positivist philosophers of science have made attempts at formalizing the logic of confirmation. These attempts are interesting (though so far unsuccessful) to researchers on *any* philosophy of science. But not only do they have nothing to do with positivist theory of meaning, they are in fact *incompatible* with it. Thus when they write about meaning, positivists tell us that 'theoretical terms' have different meanings in different theories: they formalize confirmation theory, they invariably treat theories as systems of sentences in *one* language and assume that all semantical concepts are *trans*-theoretic. Thus the positivists are engaged in formalizing *realistic* confirmation theory: not the confirmation theory (if there is one!) to which their own theory of meaning should lead."
Does Putnam realize that what positivists mean by *realistic* is idealistic and subjectivist as well as formalist, and that therefore their problem is even more involved?
126 Caution may be prudent, but not honorable. In scholastic parlance prudence is a moral virtue along with justice, temperance, and fortitude, caution being an internal part of it. The theological virtues are, of course, faith, hope and charity.
127 *Op. cit.*, p. 221.
128 *Ibid.*, p. 222.
129 This I shall call the Husserl Fallacy, based on the earlier Cartesian Fallacy of presuppositionalessness. Artistotle pointed out in the *Metaphysics* that this was impossible by the nature of human thought [as influenced by reality].
130 *Op. cit.*, p. 223.
131 According to Louis I. Bredvold, once a revered colleague, this phrase was introduced into Western thought by the Cambridge Platonists. See his *The Natural History of Sensibility* (Detroit: Wayne State Press, 1962), p. 10. This origin in idealistic philosophy is interesting because it was revived by the American and other subjectivists and positivists who generally ignore the social and physical basis of mind. In this respect see Paul Henle and Ernest Nagel.
132 *The Structure of Science*, p. 151. See also pp. 146-151.
133 *Logical Structure*, pp. 9-10.

134 *Ibid.*, p. 10.
135 *Ibid.*, p. 4. Since my shifting presuppositions have led me through Christianity, skepticism, Taoism, Zen Buddhism, Cartesianism, Process Philosophy, and Logical Positivism, I can sometimes sympathize with philosophers who hold these views, although it is perhaps not apparent in some of my polemics.
136 *Logical Structure*, p. 13.
137 See James Edwin Hansen, "An Historical Critique of Empiricism" in *Radical Currents in Contemporary Philosophy*, ed. David H. DeGrood, Dale Riepe, and John Somerville (St. Louis: Green, 1971), p. 37f. Although *what we say* the facts are is determined to some extent by our philosophy, what the facts *are* is determined by what they *are*.
138 *Logical Structure*, p. 14.
139 Robert Graves and Alan Hodge, *The Reader Over Your Shoulder* (New York: Macmillan, 1944), p 189.
140 *Logical Structure*, p. 230.
141 *Essayes*, tr. John Florio (New York: Modern Library, 1933), p. 378. (Book II, Chap. XI).
142 My italics. *Logical Structure*, p. 241.
143 See "Philosophy in India," *Philosophy in the Mid-Century*, ed. R. Klibansky (Firenze: Ia Nuova Editrice, 1959), p 296.
144 *Ibid.*, p. 299.
145 Covering the topics of time, space, grammar, irreversibility, information, cosmology, minds, machines, ESP, information theory, quantum mechanics, machines, translation, unsolvability, etc.
146 *British Journal for the Philosophy of Science*, Vol. VIII, NO. 30 (1957).
147 J.W. Dunne, *An Experiment with Time* (New York: Macmillan, 1927). In his next work, *The Universal Mind* (1938) he has "a *postulated cosmic superobserver*," a kind of Sri Aurobindo person in "pseudonow." p. 143.
148 *Ibid.*, p. 207. As. Dr. Samuel Johnson wrote in a "Preface to Shakespeare", Time is, of all modes of existence, most obsequious to the imagination; a lapse of years is as easily conceived as a passage of hours- In contemplation we easily contract the time of real actions, and therefore willingly permit it to be contracted when we only see their imitation."
149 *Ibid.*
150 "A note..," *op. cit.*, p. 156. Chari can take comfort in support from N. Razi-ud-din Siddīqui who says in "Iqbäl's Conception of Time and Space", *Iqbal as a Thinker* (Lahore: Ashraf, 1944): "Some thinkers like McTaggart have been misled in ascribing unreality to time by not keeping in mind the distinction between the serial time of the efficient ego, and the non-serial time of the pure duration of the appreciative ego.." p. 34.
151 "A Note," p. 157.
152 *British Journal of the Phiosophy of Science*, Vol. VIII, No. 32 (1958).
153 *Ibid.*, pp. 303-4.
154 It is difficult to know whether this is Symthies' or Chari's opinion. Whichever, it should be noted that we can tell when objects are moving away from us because of their becoming dimmer or smaller (the Dop-

pler Effect, red-shift in the spectrum). For quasars, according to Lovell (1968), the only shift is in dimness. If mescaline-induced images in the brain recede without getting smaller, their action is similar to quasars in this respect. The constant used in "fixing" relative motion in astronomical distances is the RR Lyrae Stars, used, says Fred Hoyle, as "standard headlights". See "The Measurement of Astronomical Distances," *Frontiers of Astronomy* (New York: Mentor, 1957), p. 150f.

155 "Hallucinations," p. 305.

156 *Ibid.*, p. 306.

157 *Estratt Rivista Methoda*, Vol. XV, Nos. 59-60 (1963).

158 *Ibid.*, p. 214.

159 *Estratt Rivista Methoda, ibid*, pp. 376-7.

160 *Dialectica*, Vol. XVII, No. 4 (1963).

161 *Ibid.*, p. 377.

162 *Ibid.*

163 In *Philosophy, Religion, and Coming Civilization*, Essays in Honor of William Ernest Hocking, ed. L.S. Rounder (The Hauge: Nijhoff, 1966), p. 383.

164 *Ibid.*

165 Quoted by Chari from Hocking's statement in *Contemporary American Philosophy*, **I.**

166 *Philosophy, Religion, and Coming Civilization*, pp. 391-2.

167 Since both think that reincarnation is a "valid possibility" [which mixes the formal with the real or existent]. Chari says that Broad is careful to add that, with all the available alleged empirical data, reincarnation must be a comparatively rare episode." It is about as rare as the appearance of rationality in the Chinese cultural revolution. See Chari, "Karma and Rebirth," *Indian Philosophical Annual*, Vol. I (1965).

168 "Human Personality....,", p. 394.

169 *Ibid*, p. 395.

170 See "Mysticism in Ethics..", *Ethics*, Vol. LXVII (1956-7), quoted in *ibid*.

171 *Ibid*.

172 *Ibid*.

173 *Quest for Truth* (New Delhi : Vohra, 1976).

174 "Crisis in Physics," in *The Concept of Freedom* (London: Lawrence and Wishart, 1977), p. 198.

175 *Quest for Truth*, p. 347.

176 *Logic of Scientific Discovery* (New York: Harper, 1968), p. 459.

177 "Philosophy of Microphysics", p. 354.

178 Lande, *Determinism and Freedom in the Age of Science* (New York: New York University Press, 1958), p. 73.

179 *Causality and Chance in Modern Physics* (London: Routledge & Kegan Paul, 1957), p. 95.

180 "Philosophy of Microphysics," p. 357.

181 *Ibid.*, p. 358.

182 *Ibid.*, p. 348.

183 *Ibid.*, p. 360. On de Broglie's view, "phenomena are no longer subject to a rigorous Determinism." *Matter and Light* (London, 1937), p. 36.

He also mentions the concept of " 'weak' causality" in *Physics and Microphysics* (New York, 1955), p. 241. De Broglie's tacit Kantianism appears on p. 239 of the same book where he writes about the former apriori character of time and space. Milić Capek, formerly of Boston University and Yale, and an old office-mate, used to regale me with arguments for non-rigid (rubberized) determinism.

184 "Arnold Toynbee's World," *Bengal Renaissance and Other Essays* (New Delhi: People's 1970). Unemployment under capitalism, including the Third World client states who are forced into the field are roughly as follows: India 120 million; Pakistan and Bangladesh 60 million; China 40 million; Latin America and Caribbean 50 million; Africa 30 million; USA and Canada 18 million; Mexico 17 milion; Western Europe 16 million; non-socialist countries in Asia 40 million; Australasia and S. Pacific 3 million. Total: 384 million. This approaches ten percent of mankind. Even Marx scarcely foresaw the horrendous deprevation concomittant wih the epidemic decline of world capitalism.

185 *Ibid.*, p. 240.

186 *Ibid.*

187 *Ibid.*

188 *Ibid.*, p. 241.

189 *Ibid.*, p. 242. A good addendum to Sarkar's analysis is that of Christopher Hill in "Time and Mr. Toynbee," *The Modern Quarterly*, New Series, Vol. II, No. 4 (Autumn 1947). According to Hill, the main defects of Toynbee as historian are confused terminology concealing tautology and platitude, denial of the importance of the productive or material base, from which it follows that he has no objective standard of progress. Ytoynbee is also prejudiced against democracy and fears the future. See p. 295.

190 *In the Sun and the Rain* (New Delhi: People's, 1972), p. 234. As John Berger says: "Realism is selective and strives towards the typical.. Naturalism in unselective....It has no basis for selection outside the present." *Art and Revolution* (New York: Pantheon, 1969), p. 50. Classicism and revolution are charged with realism.

191 *Bengal Renaissance*, p. 247.

192 *Ibid.*

193 *Ibid.*, p. 251.

194 *Ibid.*

195 *Ibid.*

196 *Ibid.*, p. 252.

197 *Ibid.*

198 *Ibid.*

199 *Man Against Society* (Chicago Regnery, 1962 [1952], p. 215.

200 *Bengal Renaissance*, p. 253.

201 *Ibid.* Lynn Rose, the American historian of cosmology and classical scholar, has noted that scientists frequently are unaware of the historical implications of their views, particularly where the scientific ideology is positivist. Rose also points out that Plato's views on the chan-

geability of the heavens are found in *Republic*, 530a-b, *Statesman* 268e-274d, *Phil.* 59a, *Timaeus* 22a-23c, 22d, 26e, 28c-j, 43a-e, *Laws* 677a ff.

202 *Ibid.*, p. 255.

203 *Ibid.*

204 (London: Methuen, 1937).

205 Much of what Stebbing writes here has been said more pointedly and with a wider perspective by Lenin in *Materialism and Empirico-Criticism* (1908), although Stebbing brings the argument up to her time by analysing the work of two British idealists in physics, James Jeans and Arthur Eddington. This phrase, "clear-thinking" and its synonsyms, much touted by the so-called philosophical analysts and linguiatic philosophers of Oxbridge and their epigoni in the U.S.A., Canada, India, Australia, South Africa, and New Zealand (not to mention Scandanavia, esp. Sweden) has the general meaning, as I understand it, of "consistent with the ideology of the speaker." However, it could, under certain circumstances mean, "consistent with certain viewpoints expressed by contemporary scientists who share our postulates". "Clarity" is invokatory and hortatory as well as mystifying among the non-initiated. "Clear-thinking" presumably requires enormous historical knowledge. At any given time it is probably insufficient for future generations reflecting on past beliefs.

206 *Bengal Renaissance*, pp. 256-7.

207 In *Society and Revolution*; Essays in Honour of Engels (New Delhi People's, 1971).

208 Sarkar quoting Engels to Schmidt, 5 August 1890 in *ibid.*, p. 122.

209 Sarkar quoting Engels to J. Bloch, 21-22 September 1890, *ibid.*, p. 123

210 Sarkar quoting Engels to H. Starkenburg, 25 January 1894, *ibid.*, p. 124.

211 *Ibid.*, p. 124-5.

212 Quoted by Dirk Struik, "Marx and Mathematics," *A Centenary of Marxism*, ed. Samuel Bernstein (New York: Science and Society, 1948), p. 181.

213 *Society and Revolution*, pp. 124-5. Probably the closest counterpart to the human hand besides the hands of the higher apes, is the trunk tip of the elephant. Elephants are not only able to scrape up salt with their prehensile member. To appreciate what the elephant can do in lumbering, transport, roadbuilding, even babysitting, see J.H. Williams, *Elephant Bill* (London: Rupert Hart-Davis, 1952). G.P. Malalasekera, formerly head of the World Buddhist Congress and Dean of Oriental Studies at the University of Ceylon, once told me in 1951 of a small female elephant which was used for baby-sitting. Her job was to keep the baby from harm by regularly pulling it with its trunk out of the reach of snakes and other possible dangers.

214 *Ibid.*, p. 126. See Marvin Harris, *The Rise of Anthropological Theory* (New York: Columbia University, 1969). Harris reasses the so-called "refutations" of Morgan. See also the work of Robert H. Lowie, *The History of Ethnological Theory* (1937), and Bernhard J. Stern, "Lewis Henry Morgan: An Appraisal of his Scientific contributions", *Historical Sociology* (New York: Citadel, 1959), and Stern's *Lewis Henry Morgan, Social Evolutionist* (1931).

215 *Aims for Oblivion* (London: Weidenfeld & Nicolson, 1966), p. 96.
216 *Society and Revolution*, p. 128.
217 *Ibid.*, pp. 134-5.
218 *Ibid.*, p. 135.
219 The Pantheon First American Edition (1965) and later in Vikas (1972).
220 Ingalls, "My Friendship with D.D. Kosambi, "*Indian Society: Historical Probings*, p. 26.
221 *Ancient India*, p. 135. This statement was paralleled by Vladimir Mshvenieradze of the Department of Philosophy of Moscow State University in a lecture at the State University of New York at Buffalo, 12 February 1971. He said, "I would rather have knowledge than theory of knowledge". But how would he know that he had the knowledge without a theory of knowledge?
222 Indian Feudal Trade Charters, *Journal Econ. Soc. Hist. Orient.* 59, vol. 2, pp 281-93; "At the cross-road, Mother goddess cult-sites in ancient India," *J. Asiatic.*, 60, p. 17-31; "Social and economic aspects of the Behagavad Gita," *J. Econ. Soc. Hist. Orient.* 61, 4, pp. 198-244; "An Inscription at Palasdev of Saka 1079," *J. Asiatic Soc.* (Bombay: New Series) 1963, pp. 69-73; "Challenges of evolution," *Seminar* 82 Je' 66 pp. 13-19; "Bhartri-Hari, the Poet." Bharatiya Vidya Bhavana: Bombay 1959; "Continuous Groups and Two Theorems of Euler, *Mathematics Student*, Madrs, 1934; "On Differential Equations with the Group Property," S.V. & Co.: Madras 1932; "The working class in the Amartakosa," *J. Orient. Res.* 24 (1-4), pp. 57-69 [Armarakosa is the most ancient dictionary of Sanskrit, as a source for sociological research]; "Megaliths in the Poona District, *Man* 62: pp. 65-67; Expanding Universe, *Nature* [30 : 507, Oct. 1, 1932; "Caste and Class in India, *Science and Society* 8, 3: pp. 243-9 1944; "Geometric Method in Mathematical Statistics," *Am. Math. Mo.* 51: p. 382-9, Aug. 1944; Staple grains in the Western Deccan," *Man* 63 Aug. 1963, pp. 130-1; "Lie Rings in Outer Space," *Nat. Acad. Sci. Proc.* 35: pp. 389-94; "Prehistoric rock engravings near Poona," *Man* 63, Apr. 1963, pp. 57-8;"What constitutes Indian history," *Bhandarkar Orient. Res Inst.* 35 (1-4) 54 : pp. 194-201.
223 *Ancient India*, p. 10.
224 *Ibid.* p. 12.
225 *Ibid.*, p. 16.
226 Pareto, *The Mind and Society*, trans. A. Bonglorno (New York: Harcourt, Brace, 1935) II, 511.
227 *Ibid.*, p. 519.
228 If it is instinctual.
229 *Ancient India,* p. 47.
230 (Bombay: Popular Prakashan, 1962).
231 See his account of Mimamsa ritual in Indian Philosophy (New Delhi: People's, 1969).
232 As in Aristotle's *Nichomachian Ethics* with its doctrine of the Mean. Also in Stoicism and Epicureanism.
233 *What is Living and What is Dead in Indian Philosophy* (New Delhi; People's), p. 112.

234 *Ibid.*, ii 5.2.27, cited in *What is Living*, p. 113.
235 For detailed discussion of law in this connection see his *Science and Society in Ancient Indian* (Calcutta: Research India Publications, 1977), pp. 212-360.
236 *Ibid.*, p. 115.
237 *Ibid.*, p. 117.
238 *Ibid.*
239 *Ibid.*, p. 118.
240 *Ibid.*, p. 126.
241 *Ibid.*, p. 127.
242 *Ibid.*, p. 135
243 *Ibid.*, p. 157.
244
245 *Ibid.*, pp. 168-9.
246 *Ibid.*, p. 196f.
247 *Ibid.*, p. 198.
248 *Ibid.*, p. 220.
249 Rahman propably means over-abstract or metaphysical or idealistic by "over-philosophizing". Rahman correctly refers to the views of J.B.S. Haldane as supplementing other views of the science of science. See especially his *The Causes of Evolution* (1932), *The Marxist Philosophy and the Sciences* (1946), *Possible Worlds*, (1928), *What is Life* (1947), & Ronald W. Clark, *J. B.S. : The Life and Work of J.B.S. Haldane* (1969.) Ch. 12 deals with Haldane's work in India. For Haldane's interconnections with Bernal, Hyman Levy and Needham see Gary Werskey *The Visible College* (NY: Holt, Rinehard & Winston, 1979).
250 "The Approach to Science," *Society and Revolution*, p. 173.
251 See J.D. Bernal, *The Origin of Life* (London: Weidenfeld and Nicolson, 1967), where he claims that the first half of the twentieth century was dominated by a revolution in physics, whereas the second half is dominated by the revolution in biology. Perhaps we are having both today.
252 "The Approach to Science", p. 175.
253 *Ibid.*, p. 176.
254 See Gesellschaft Fūr Deutsch-Sowjetische Freundshaft, *Uber Formale Logic und Dialektik* (Berlin: Verlag Kultur and Fortschriftt, 1954). See also E.V. Ilyenkov, *Dialectical Logic* (Moscow: Progress, 1977), E.V. Ilyenkov, *The Dialectics of the Abstract and the Concrete in Marx's Capital* (1982),
255 See various issue of *Soviet Studies in Philosophy*, ed. John Somerville in logic and philosophy of science.
256 "Approach to Science," pp. 179-80.
257 *Ibid.*, p. 180.
258 In *Societv and Revolution*, Essays in Honour of Engels (New Delhi: 1971).
259 *Principles of the Theory of Probability* (Chicago: Universitv Press, 1939), p. 3.
260 *Ibid.*, p. 65.

261 *Pragmatism versus Marxism* (New York: Pathfinder, 1975), p. 279.

262 See also Harry K. Wells, *Pragmatism: the Philosophy of American Imperialism* (London: Lawrence & Wishart, 1954). According to Wells, "Dewey's 'continuity of nature and experience,' like James' 'radical empiricism' opens wide the way to obscurantism." (p. 143) He also says: "No member of society, including the worker, can avoid the corrupting influence of pragmatism. It enters his consciousness through a variety of channels. It invades his home through the radio, the press, the education of his children. It penetrates his union through the misleaders and hired agents. It confronts him at work through the entire process of production for profit in which skilled labor is subordinate to 'getting the job done' somehow and as fast as possible. It poisons him through the cynical corruption and graft inherent in bourgeois politics....The vicious ideology of white supremacy, fostered by the imperialist oppressors, is a prime source of the pragmatic virus, for it leads to unprincipled practice. It leads to expedient adjustment to the special oppression of the Negro people, and thus undermines principle in all aspects of life. The ideology of male supremacy is likewise a major source of the pragmatic outlook as it penetrates the working class in the home and in the factory. Here again there is compromise with principle and adjustment to the special oppression of women, which further heightens the corruption of theory and the transformation of practice into opportunistic expediency." pp. 202-3. In short, pragmatism is incompatible with science and the ethics of equity.

263 Transactionalists, like Rollo Handy of the American Institute for Economic Research, defend Dewey on the ground that he was among the best of American social thinkers (in academic philosophy) in his time. John Somerville, once a student of Dewey, but today no follower of his, tells me in a recent letter that Dewey was the only liberal spot, the only bright light, at Columbia University in the 1920's and 1930's. I can understand these arguments. Nevertheless the serious weaknesses of his position cannot longer be papered over. His influence has been seriously eroded since Americans began reading histodiamat literature.

In contemporary Chinese social thought, Dewey is seen as a defender of the Hu Shih (and hence, Chiang Kai Shek) forces, and an apologist for bourgeois philosophy. See Hou Wai-lu et al., *A Short History of Chinese Philosophy*, trans. Wang Cheng-chung (Peiking: Foreign Languages Press, 1959), p. 113. See also "Hu Shih" in *Sources of Chinese Tradition* (New York: Columbia University, 1964), II, 169-72. While in China Dewey sent secret reports to the State Department of the United States concerning the Chinese "situation" so was acting as an intelligence agent. He tried to persuade the Chinese youth that socialism was not the answer to their problems, but some form of liberalism that "worked so well" in the United States. Dewey reached the peak of his progressiveness in the 1920's (while he was already being quoted by Hu Shih against socialism), after which he grew increasingly conservative. The touchstones of reformism, according to J.D. Bernal, are abandonment of socialism, emphasis upon biological evolution as opposed to scientific revolution, and the acceptance of marginalist economics which has the merit of making exploitation disappear into the woodwork. See Bernal, *Science in History*, IV, 1112. Says Bernal of Dewey: "For all his "advanced"thought, his views are "scientifically empty and morally bankrupt [and]. .quite inadequate to resist the openly reactionary and obscurantist trends of Americanism." *Ibid.*, 1st ed. p. 754.

264 Bernal, *Science in History*, 3rd ed., IV, pp. 1113-38. See Gary Werskey, *The Visible College* (New York: Holt, Rinehart, Winston, 1978), for an account of Cambridge University science in Bernal's day.

265 Bose, *op. cit.*, p. 186, quoting from letter to P.L. Lavrov (Nov. 1875).

266 *Ibid.*

267 Hou Wai-lu, *op. cit.*, p. 33.

268 *Anti-Duhring* (Moscow: 1969), pp. 18-19, quoted by Bose, *op. cit.*, p. 187.

269 *Ibid.*

270 (London: Weidenfeld & Nicolson, 1967), p. 4.

271 *Introduction a L'etude de la Medicine Experimentale* (Paris: Delagrave, 1920), p. 118.

272 Bose, *op. cit.*, pp. 189-90.

273 *Ibid.*, p. 190. For detailed accounts of *energeia* and *dynamis* as they were construed by Aristotle, see C.H. Chen, *Sophia the Science Aristotle Sought* (Hildesheim: Olms, 1976).

*See f.n. 184, p. 205.

NOTES FOR CHAPTER III

1 *The Awakening of the Faith*, tr. Timothy Richard (Sanghai, 1907), quoted in *The Tibetan Book of the Dead*, ed. & tr. W.Y. Evans-Wentz (New York: Causeway, 1973), p. 226.

2 Feibleman, "Full Concreteness and the Re-Materialization of Matter," *Diogenese*, 60 (Winter, 1967), p. 50. See also his *The New Materialism* (1970), p. 55. Errol E. Harris says in *The Foundations of Metaphysics in Science* (New York: Humanities, 1965), p. 37 that "Modern physics has not abolished matter, or replaced it by 'spirit', it has given a fundamentally new account of it; and mechanism has not given way to 'vitalism' or anything of that sort...." That is true, but in some circles mechanism has been replaced by dialectical materialism.

3 Dirk Struik, "Mathematics," *Philosophy for the Future*, ed. R.W. Sellars, V.J. McGill and Marvin Farber (New York: Macmillan, 1949), p. 138.

4 Hermann Weyl, *Philosophy of Mathematics and Natural Science* (Princeton University Press, 1949), pp. 65-66. Poincaré was not satisfied with subjectivism, for he says: "The possibility of mathematical science seems an insoluble contradiction. If this science is only deductive in appearance whence comes it that absolute rigor which nobody thinks of contesting? If on the contrary all propositions which it establishes can be obtained from one another by rules of formal logic, why does not mathematics reduce itself to an immense tautology? *La Science et l'hypotheses* (Paris, 1918), quoted by Struik, *Concise History of Mathematics*, pp. 145-6. Brouwer thought of mathematics as "a free creation independent of experience." Although Weyl says of Brouwer that he "represent idealism by demanding the reduction of all truth to the intuitive given," he himself thought that "we are satisfied with the symbol...." It was A.S. Eddington in *Science and the Unseen World* (New York: Macmillan, 1931), as Plumian Professor of Astronomy at Cambridge University, while presenting lectures on science and religion to fellow Quakers who said: "We have learnt that exploration of the external world by the methods of physical science leads not to a concrete reality but to a shadow world of symbols...." p. 73. Advaita has never expressed it better.

Eddington once remarked that he thought that all male materialists "must think of their wives as differential equations." J.B.S. Haldane replied that "Men have fallen in love with statues and pictures. I find it far easier to imagine a man falling in love with a differential equation, and I am inclined to think that some mathematicians have done so." See Gary Werskey, *The Visible College* (New York, 1979), p. 85. It was Renoir's view that the value of non-physical subjects in art approached zero.

5 Preface to *Dialectics of Nature* tr. Clemens Dutt (London, 1946).

6 *Contemporary Theories of Knowledge* (New York: Ronald, 1961), p. 358.

7 "Preface," Engels, *Dialectics of Nature*, XIV-XV.

8 (Cambridge: Harvard, 1949), pp. 193-4.

9 *Revolutionary World*, Vol. 6 (1974), p. 75.

10 My italics—DR. Frank *op. cit.*, p. 197.
11 *Ibid.*
12 Besides his *Philosophy of Artihmetic* (1891), "Psychological Studies in Elementary Logic" (1894), *Logic Survey* (1895-9), Logical Investigations (1900-04), rev. 1913), *Experience and Judgment* (1939), Kneale in *The Development of Logic* (Oxford, 1968) does not mention Husserl nor does K.O. May, *Bibliography and Research Manual of the History of Mathematics* (Toronto, 1973). Perhaps they regard most of what Husserl himself believed to be mathematics was introspective psychology.
13 "Experience and Subjectivism," *Philosophy for the Future*, p. 630.
14 *Ibid.*
15 *An Examination of Logical Positivism*, [(1963], 1960).
16 *In Defense of Philosophy* (London, 1950).
17 "Science and Pseudo-Science: Aspects of the Class Struggle in the Sphere of Philosophy," *Marxism today*, Vol. 16, No. 9 (Sept. 1972), p. 271.
18 *Ibid.*
19 See R.G. Collingwood, *Faith and Reason*, ed. Lionel Rubinoff, (Chicago, 1968), 88-92.
20 "Science and Pseudo-Science," p. 270.
21 *Ibid.*
22 *Ibid.*, p. 271.
23 *Ibid.*
24 *Ibid.*
25 *Ibid.*, pp. 271-2.
26 *Ibid.*, p. 272.
27 See Maurice Cornforth, *The Open Society and the Open Philosophy*, (London, 1968).
28 Contenders are: economic theories of value, consequences of political interventionism, parameters of capitalist monopoly operations, origins and significance of equality, social engineering, pragmatic imperialism. An example of Popper's relation to fideism may be seen in a letter to the editor of *Scientific American*, Vol. 235, No. 1 (July 1976), p. 6: the writer, Dr. Robert E. Kofahl, Science Coordinator, Creation-Science Research Center, San Diego, California, says: "....evolutionary explanations of origins also fall outside the realm of science strictly defined, because they are *unfalsifiable*. And creationists have very respectable support in this position, for example from science logician Karl Popper... " [my italics—DR].
29 Gutzmore, *op. cit.*, p. 272.
30 *Ibid.*
31 *Idid.*, pp. 272-73.
32 *Ibid.*, p. 273. Gutzmore makes a fourth objection which hinges on, he claims, whether Popper and Lakatos understand Descartes' distinction concerning instantiation theory and "talk about" syntactically metaphysical propositions. Since he does not develop this point I shall only make this passing reference to it. See *Ibid.*, and 273n.
33 *Ibid.*, p. 274.
34 *Ibid.*

35 *Ibid.*
36 *Ibid.*, p. 268.
37 *Ibid.*, p. 274.
38 *Ibid*. Objective idealism appeared long before Descartes. Subjective idealism developed concommitantly with bourgeois power over the aristocrats royalists, and feudal landowners. It was the power of money (symbolic) over landed property.
39 (The Hague: Nijoff, 1971), p. 3.
40 *Structure of Scientific Revolutions*, 2d ed. (Chicago: University Press, 1970), p. 207.
41 *Ibid.*, pp. 180-5.
42 *Introduction to the History of Science*, I, 19.
43 *Op. cit.*, p. 268.
44 (Cambridge: Harvard, 1959), pp. 48-9.
45 *Structure*, p. 164.
46 See Wladyslaw Tatarkiewicz, *Nineteenth Century Philosophy*, tr. C.A. Kisiel (Belmont, California: Wadsworth, 1973).
47 "Yajnavalkya and the Philosophy of Fictions," *The Memorial Volume* (Poona : Oriental, 1973), p. 274. Vaihinger's subjectivist pathology has seldom been equalled in the West except by such phenomenologists as Maurice Merleau-Ponty. See his *The Visible and the Invisible, Followed by Working Notes*, tr. Alphonso Lingis (Evanston: North-western University, 1968), and his *Phenomology of Perception*, tr. Colin Smith (London: Routledge & b Kegan Paul, 1965).
48 *Structure*, pp. 206-7. The very concept of "puzzle-solving" as an activity having little or no connection with the social history of man is subjectivist, positivist, and metaphysical in Cohen's sense. The comparison of Einstein and Aristotle is a promiscuous analogizing that reminds me of a remark by Groucho Marx in the movie *Animal Crackers* where Groucho alleges that grapefruit tastes more like cranberries than cantaloupe does, or something similarly fatuous and amusing.
49 William Appleman Williams in *The Great Evasion* (Chicago: Quadrangle, 1964), shows how the evasions of histodiamat has adversely affected American foreign policy and political economy. It seems likely also that it has impoverished American philosophy of science.
50 *Marxism Today*, Vol. 19, No. 8 (August 1975), p. 253. In a footnote on the same page, Thompson says that positivism rules out all real science, referring for proof to the science and philosophy to be found in *The Philosophy of Rudolf Carnap*, ed. P.A. Schlipp.
51 *Structure*, p. 170.
52 *Ibid*.
53 (New York: Vintage, 1959).
54 DeGrood, review of *Critism and the Growth of Knowledge: Proceedings of the International Colloquium in the Philosophy of Science*, ed. Imre Lakatos and Alan Musgrave (1970) in *Science and Society*, Vol. XXXVI, No. 3 (Fall 1972).
55 "Dialectics and Scientific Revolutions," *Science and Scciety*, Vol. XXXVII, No. 3 (Fall 1973). Not all subjectivists are idealists in all

respects. Remy de Gourmont (1858-1915) for example, believed thoughts were material yet the world is reduced to the measure of the mind. His outlook has been called "sensual intellectualism". He and his fellow Symbolists opposed the naturalism of Comte, Littre, Renan, Taine and Zola. Yet these Symbolists were, as Anatole France pointed out, hyper-individualistic, obscurantist and arrogant. One might wonder whether any philosophical or critical school ever had a monopoly on arrogance. See Glenn S. Burne, *Remy de Gourmont, His Ideas and Influence in England and America* (Carbondale, 1963).

56 "Notes on 'Dialectics of Nature'."

57 *Ibid.*, p. 254.

58 *Op. cit.*, p. 379.

59 *Structure*, p. 171.

60 *Ibid.*

61 *Ibid.*

62 *Ibid.*

63 *Ibid.*

64 See Imre Lakatos, *Criticism and the Growth of Knowledge*.., pp. 179-80.

65 *Op. cit.*, p. 332.

66 *Structure*, p. 162.

67 *Ibid.*, p. 170.

68 *Ibid.*

69 *Ibid.*, p. 173.

70 *Ibid.*

71 *Ibid.*

72 E.J. Hobsbaum, "On Marx's Pre-Capitalist Economic Formation," *Enquiry* (New Series), Vol. III, No. 2 (Monsoon 1969), p. 4.

*

NOTES FOR CHAPTER IV

1 "Il descend, reveille, l'autre cote du reve," a line from Hugo's "Lu Bouche d'ombre," *Contemplation.*
2 But see footnote 87 for a view of Kant's progressivism.
3 In this account the subjectivist is anyone essentially agreeing with Kant's distinction that reality is unknowable and that what we know is our experience—(often of the unknowable).
4 *Naturalism and Subjectivism* (Springfield, Illinois: Thomas, 1959), p. 5.
5 *Ibid.*, p. 8.
6 An idealist is one who believes in the primacy (in time or history) or exclusiveness of consciousness over matter or nature.
7 J.D. Bernal, "The Social Sciences in History," *Science in History* (New York: Cameron, 1954), p. 721f.
8 *Ibid.*, p. 724.
9 *Ibid.*
10 The excesses of hypostatized reason without scientific justification was one avenue to revived empiricism and subjectivism in the nineteenth and twentieth centuries.
11 *Republic oj the West Order and Progress*, tr. J.H. Bridges [Paris, 1848] (London: Routledge, n.d.), p. 55.
12 *Ibid.*, p. 355.
13 Quoted by Bernal, *op. cit.*, p. 732.
14 A neologism combined of historical materialism and dialectical materialism.
15 See especially the work of F.Y. Edgeworth and Vilfredo Pareto where a cloud of mathematics covers the anti-scientific artillery. Evidently even orthodox ecomonists are not overwhelmed by the significance of Edgeworth's mathematizations. Schumpeter says, for example, "I am unable to work up any enthusiasm for his manner of introducting apriori [idealistic] probabilities into purely theoretical reasoning [idealistic]." *History of Economic Analysis* (New York: Oxford, 1968), p. 961. We might add that we are unable to work up any enthusiasm for Schumpeter's use of "purelv". See the section on INDEPENDENCE AND PURITY, p.
16 "In Defence of the Marxist Impact," *The Impact of Marxist on Indian Life and Literature* (Mysore: Institute of Kannada Studies, 1971), p. 4.
17 *Ibid.*, p. 8.
18 The more impressive crises occurred in 1808-11, 1815-21, 1825-27, 1829, 1830-33, 1837-38, 1840, 1843, 1846-49, 1857, 1862-63, 1870, 1883, 1890-91, 1907-10, 1929-41, 1949-51, 1960-61, 1969 to the present. Under capitalism they have all had international impact.
19 The best statements of this are to be found in Lenin's *Materialism and Empirio-Criticism.* Charistopher Caudwell's *Crisis in Physics,* and L. Susan Stebbing's *Philosophy and the Physicists.* Lenin and Caudwell understand the ideological implications of subjectivism in physics.
20 Anthropologists had some advantage over other social scientists, especially philosophers, in retaining an objective outlook because they

were directly aware in their field work of the importance of the productive process for survival, saw the importance of technology at a clear and simple level and were more aware of imperialist exploitation.

21 Or systematically inaccurate, whichever made for maximization of profits.

22 Joseph Schumpeter (1883-1950) seems to be an exception.

23 Anthropologists and psychologists tend to be more calm because their entire existence is not threatened.

24 *A History of Economic Thought*, fourth edition (London: Faber and Faber, 1973), p. 289.

25 Apologetics on this score are seen in the treatment of use value versus exchange value by Frank Knight who notes as recently as 1950 in his "Introduction" to Menger's *Principles of Economics* (Glencoe: Free Press, 1950), p. 15 that: "It required about a century of waiting.. to point out that.. use value does correspond to exchange value....The paradox was, of course, resolved by noting the self-evident fact [self-evident fact that had been missed for one hundred years!] that the use value in question is not the difference it makes whether one has an ordinary ration of the good or has none, but the significance of having a little more or a little less." This touch of inscrutability was part of the Austrian legacy in the United States.

26 Maurice Cornforth, *The Open Philosophy and the Open Society* (London: Lawrence and Wishart, 1968), p. 197. The attempt to separate price and commodities from other social relations besides subjective wants doomed orthodox econonmics to permanent adolescence.

27 Albven Barkley, former Vice President of the United States, said that an economist was "a gu g with a Phi Beta Kappa key on one end of his watch chain and no watch on the othr end."

28 "I myself once spoke of economic theory..as applied psychology." said Friedrich von Wieser, *Social Economics*, tr. A.F. Hinrichs (New York: Adelphi, 1927), p. 3. This is correlative to Einstein's remark that physics is mostly metaphysics.

29 *New York Times* (June 4, 1974).

30 No other subjectivist has been able to compete with Kant's inscrutable technical apparatus. Coming closest are the phenomenologists who have produced among other concepts: *Erfahrungsgegenstandlichkeiten* [italics—DR] (ideal objects of experience).

31 *A History of Philosophy*, tr. James H. Tufts (New York: Macmillan, 1901).

32 *History of Philosophy*, tr. G.S. Morris (New York: Scribiner's Sons, 1893).

33 Rev. ed. (New York: Holt, 1948), p. 512. This wins the prize as the most mealy-mouted conclusion of all the histories I looked at. It appears to be the result of California naturalism combined with rotarianism.

34 *History of Philosophy*, tr. S. Applebaum and C.C. Strowbridge (New York: Dover, 1967), p. 468.

35 *A History of Russian Philosophy*, tr. G.L. Kline (New York: Columbia University, 1953), II, 924.

36 *Essays in Positive Economics* (Chicago: Phoenix, 1966), p. 21n. I will not dignify this by calling it metaphysics.

37 Knowledge of states of mind in itself is not incompatible with knowledge of reality.

38 The Viennese, Ludwig von Mises, has an occasional lapse at which time a kind of refreshing modesty appears. He says that.."the relationship of economics to psychology is still problematical. The position due Gossen's law of the satiation of wants yet remains to be clarified [approaching 125 years later)]." *Epistemological Problems of Economics*, tr. George Reisman (Princeton: Van Nostrand, 1960), p. 152.

39 "The lines of division between future, past, and present consumption remain as blurred in Boehm-Bawerk's formulation as they do in Menger's and Wieser's presentation of the time factor in relation to consumption." Paul Crosser, *Prolegomena to All Future Metaeconcomics*, (St. Louis: Green, 1974) p. 109. We will remember that Kant thought that space and time were subjective, which means that when the astronauts take a space walk, they are walking in their consciousnesses.

40 But really reserved for entrepreneurs who by "abstinence" or as Marshall preferred to call it, "waiting," set the conditions for *future* pleasure—which can never come. Is it *within* human nature for everyone to try to maximize pleasure? If so, it surely follows that given half a chance the poor worker would be just as grasping as the rich capitalist which only proves that human nature can't be changed—even under bureaucractic communism! Social Darwinism was employed to push such innatist views with an assist from Fabianism.

41 Crosser, *op. cit.*, pp. 113f. The notion of a qualitative minimum may have come from Averroes (1126-1198), (Ibn Rushd), the commentator on Aristotle. See Andrew G. Van Melsen, *From Atomos to Atom* (New York, 1960), p. 59.

42 *Ibid.*

43 Schutz, *The Phenomenology of the Social World*, tr. G. Walsh, and F. Lehnert (Evanston: Northwestern, 1967), p. 52.

44 *Ibid.*, p. 78.

45 *The Journals of Andre Gide*, tr. J. O'Brien (New York: Knopf, 1951), IV, 139. Paralled to this subjectivism in economics is subjectivism in art. Maurice Denis, French critic held that Nature for the artist is "a state of his own subjectivity." "De Gauguin et de Van Gogh au classicism," 1909, *Theories*, p. 276. For Gauguin, the work of art "is a mirror wherein is reflected the state of soul of the artist." J. de Rotonchamp, *Paul Gauguin*, p. 152. The content, the subject matter of the painting as a picture of something becomes secondary to the subjective intention of the artist. Objective judgment of art objects is replaced by the psychology of the artistic soul. A new rise of religious sentiment followed, and with it a renaissance of Roman Catholic philosophy (Blondel, Peguy, Laberthonniere).

46 *Economic Fictions, A Critique of Subjectivist Economic theory* (New York: Philosophical Library, 1957), p. xi.

47 See von Thunen, *Der Isolierte Staat* (1930), pp. 11-12.
48 *Essays in Positive Economics*, p. 15.
49 *Ibid.*, p. 15.
50 *Ibid.*, pp. 17-18.
51 *Ibid.*, p. 19.
52 The disappearance of ontology also affects the sense of causality which the subjectivists after Hume identified with succession. Getting the cause and effect turned around is endemic to idealism. P.F. Willer says in "Philosophy and the Revolution," *The Cambridge Modern History* (New York: Macmillan, 1928), VIII, 1: "Intellectual anarchy prepared the way for social anarchy." Alvin H. Hansen, the American economist, contributes the following to this trend: "A country with a vast capacity to save is not in great danger of inflation." *Economic Policy and Full Employment* (New York: McGraw-Hill, 1947), p. 6. It might be pointed out also that *countries* never save anything, especially the United States, which is the most wasteful of all capitalist states, because it is the most advanced and the most hypostatized !
53 *Essays in Positive Economics*, p. 20.
54 *Ibid.*, p. 24.
55 *Ibid.*, p. 7.
56 Marx, an economist who was really enthusiastic about theory, said:, "..theory itself becomes a material force when it has seized the masses." "The Critique of Hegel's Philosophy of Right," tr. Bottomore, *Karl Marx, Early Writings*, p. 52.
57 *Ibid.*, p. 33.
58 Is this merely Milton Friedman's reliquary of Comte-ian sacred remains? A materialist informant claims that psychological studies show that subjectivists are usually far behind in their letter writing which induces in them a feeling of displeasure when faced by the word "correspondence." At the same time subjectivists tend to be near the front of the cafeteria line. Materialists tend to be gross in their preference for "correspondence" over "coherence."
59 *Ibid.*, p. 39.
60 *Ibid.*, p. 35.
61 *Ibid.*, p. 30.
62 My italics. *Ibid.*, p. 42.
63 Irene Brennan, "The concept of Reflection," *Marxism Today*, Vol. 18, No. 4 (April 1974), p. 126. This discussion is comprehensive, timely, and remarkably clear.
64 V.I. Lenin *Philosophical Notebooks* tr. Clemens Dutt. Collected Works, (London: Lawrence and Wishart, n.d.), XXXVIII, 171.
65 Mills, *The Sociological Imagination*, p. 67.
66 *Ibid.*, p. 64.
67 *Economic Fictions*, pp. xvii-xviii.
68 *Social Economics*, p. 4. My italics.
69 *Ibid.*, p. 10.
70 Marvin Farber, *Phenomenology and Existence* (New York: Harper, 1967), p. 30.

71 (New York: Oxford, 1968).
72 Schumpeter, *op. cit.*, p. 559n.
73 Admittedly a great deal more analysis is required before we can understand the full meaning of "purity" in orthodox social science. It will require, I am afraid, a history of purification and its derivations from ancient society to the present. That it will have intimate connection with puritanism, fideism, anti-materialism and eliteism, I cannot doubt.
74 (London: Weidenfeld & Nicolson, 1964), pp. 183-4.
75 Needham, *Time, The Refreshing River* (New York: Macmillan, 1943), p. 87.
76 *Ibid.*, pp. 87-8.
77 Quoted by Needham, *ibid.*, p. 88 from Auden's "New Year Letter" (1941), p. 34.
78 *The Watchdogs*, tr. Paul Fittingoff (New York: Monthly Review, 1971), p. 81. Trotsky observed that "there is no creature more disgusting than a petty bourgeois engaged in primary accumulation." What about a petty bourgeois not engaged in primary acculmulation?
79 Borden P. Bowne, *Metaphysics* (New York: Harper & Brothers, 1898), p. 427.
80 Unless his furnace goes out in the winter or his reading light goes out at night.
81 *Watchdogs*, pp. 81-2, my italics.
82 *The Private Journal of Henri Frederic Amiel* [1821-88], tr. Van Wyck and Charles Van Wyck Brooks (New York: Macmillan, 1935), p. 458.
83 *The Sociological Imagination* (London: Oxford, 1967), p. 67.
84 *Ibid.*
85 *The World of C. Wright Mills* (New York: Marzani & Munsell, 1960), p. 102. Aptheker also notes on the same page that the operationalists Polykarp, Kusch, Bridgman, and W.S. Beck believe that there is *no* scientific method. The scientist just "does his utmost" with no holds **barred !**
86 *Op. cit.*, p 75.
87 Kant himself was favorable to the American Revolution, to technological advances, to the view that antagonism plays a role in the historical process, and to eternal peace.
88 *E.G.* Christian socialism, co-operatism, utopianism.
89 I was awakened from my dogmatic slumbers regarding the "dead issue" of fideism by the late Roman Ingarden who in "About the Motives Which Led Husserl to Transcendental Idealism," *Phenomenology and Natural Existance*, ed. D. Riepe (Albany: State University of New York Press, 1973), p. 115. Ingarden points out that Husserl hoped to find liberation from the burden of awareness of the "badness of the world" through faith founded in the transcendental.
90 It might appear odd to associate anti-intellectualism with subjectivists, particularly since no one has appeared more ingenious in developing intellectual props for fideism, anti-materialism, and anti-vulgarism. Deeper than this, however, is the underlying desire to leave the world unexplained and instead toy with appearances that will avoid forever facing-up to the political and scientific issues. The recent popularity of game theory is another indication of this. Much of cybernetics "theory" is also horse-play to avoid resolution of basic issues. Computerization of subjectivism (garbage-in-garbage out) is the latest diversion.

INDEX

N

O